MISSING LINKS IN LABOUR GEOGRAPHY

T0300291

Missing Links in Labour Geography

Edited by

ANN CECILIE BERGENE
Work Research Institute, Norway

SYLVI B. ENDRESEN
University of Oslo, Norway

HEGE MERETE KNUTSEN
University of Oslo, Norway

LONDON AND NEW YORK

First published 2010 by Ashgate Publishing

Published 2016 by Routledge
2 Park Square, Milton Park, Abingdon, Oxfordshire OX14 4RN
711 Third Avenue, New York, NY 10017, USA

First issued in paperback 2016

Routledge is an imprint of the Taylor & Francis Group, an informa business

Copyright © Ann Cecilie Bergene, Sylvi B. Endresen and Hege Merete Knutsen 2010

Ann Cecilie Bergene, Sylvi B. Endresen and Hege Merete Knutsen have asserted their right under the Copyright, Designs and Patents Act, 1988, to be identified as the editors of this work.

All rights reserved. No part of this book may be reprinted or reproduced or utilised in any form or by any electronic, mechanical, or other means, now known or hereafter invented, including photocopying and recording, or in any information storage or retrieval system, without permission in writing from the publishers.

Notice:
Product or corporate names may be trademarks or registered trademarks, and are used only for identification and explanation without intent to infringe.

British Library Cataloguing in Publication Data
Missing links in labour geography. -- (The dynamics of
economic space)
1. Labour market. 2. Labour unions. 3. Labour unions--
Political activity. 4. International labour activities.
5. Economic geography.
I. Series II. Bergene, Ann Cecilie. III. Endresen, Sylvi.
IV. Knutsen, Hege Merete.
331.1'2-dc22

Library of Congress Cataloging-in-Publication Data
Bergene, Ann Cecilie.
Missing links in labour geography / by Ann Cecilie Bergene, Sylvi B. Endresen, and Hege Merete Knutsen.
p. cm. -- (The dynamics of economic space)
Includes index.
ISBN 978-0-7546-7798-7 (hardback) -- ISBN 978-0-7546-9554-7 (ebook)
1. Human geography. 2. Economic geography. 3. Labour movement. 4. Labour unions.
I. Endresen, Sylvi B. II. Knutsen, Hege Merete. III. Title.
GF50.B47 2010
331--dc22

2010014399

ISBN 13: 978-1-138-26989-7 (pbk)
ISBN 13: 978-0-7546-7798-7 (hbk)

Contents

PART III POLITICS OF LABOUR

PART IV LABOUR AND STRATEGIES OF CAPITAL

PART V CONCLUSION

List of Tables

List of Contributors

Dr. Gunilla Andrae is retired Reader from the Department of Human Geography at Stockholm University.

Dr. Björn Beckman is Professor Emeritus at the Department of Political Science at Stockholm University.

Dr. Niels Beerepoot is Lecturer and Researcher at the Department of Geography, Planning and International Development Studies of the University of Amsterdam.

Dr. Ann Cecilie Bergene is a Senior Researcher at the Work Research Institute in Oslo, Norway.

Dr. John Bryson is Professor of Enterprise and Economic Geography at the University of Birmingham (UK).

Dr. Neil Coe is Reader in Economic Geography at the University of Manchester.

Dr. Andrew Cumbers is Senior Lecturer at the Department of Geographical and Earth Sciences, University of Glasgow.

Dr. Jamie Doucette is a graduate of the Department of Geography, University of British Columbia.

Dr. Sylvi B. Endresen is Associate Professor at the Department of Sociology and Human Geography at the University of Oslo.

Dr. Martina Fuchs is Professor at the Department of Economic and Social Geography, University of Cologne, Faculty of Management, Economics and Social Sciences.

Eva Hansson is a PhD Candidate at the Department of Political Science, Stockholm University.

Dr. Andrew Herod is Professor of Geography, Adjunct Professor of International Affairs, and Adjunct Professor of Anthropology, University of Georgia, USA. He is also an elected official, being a member of the government of Athens-Clarke County, Georgia.

Herbert Jauch is Head of Research and Education at Labour Resource and Research Institute in Windhoek, Namibia.

Dr. David Jordhus-Lier is Senior Researcher at the Norwegian Institute for Urban and Regional Research.

Dr. Hege Merete Knutsen is Professor in Human Geography at the Department of Sociology and Human Geography, University of Oslo.

Ola Anders Magnusson holds a Master's Degree in Human Geography. Department of Sociology and Human Geography, University of Oslo.

Dorit Meyer is a PhD Candidate at the Department of Economic and Social Geography, University of Cologne, Faculty of Management, Economics and Social Sciences.

Dr. Paul Routledge is Reader at the Department of Geographical and Earth Sciences, University of Glasgow.

Rebecca Ryland is an ESRC-funded PhD Candidate at the Department of Geography, University of Liverpool.

Dr. Mike Taylor is Professor of Human Geography at the University of Birmingham (UK).

Dr. Steven Tufts is Assistant Professor in the Department of Geography at York University.

PART I
Introduction

Chapter 1
Re-engaging with Agency in Labour Geography

Ann Cecilie Bergene, Sylvi B. Endresen and Hege Merete Knutsen

The erosion of the Keynesian welfare state and the intensification of neoliberal economic globalization have inevitably brought issues concerning the welfare and working conditions of labour to the forefront both in the global north and the global south. The era of expanded reproduction, which was characterized by the protection of infant industries producing for an internal mass market and the inclusion of trade unions as important bargaining partners has, since the crisis of the 1970s, been supplanted by a search for profits through accumulating by dispossessing the masses of public goods and benefits (Harvey 2003), and a quest for gains in productivity and for increases in the rate of exploitation to stave off the crisis of Fordism (Lipietz 1982). The new accumulation regime went hand in hand with the rise to hegemony of the neoliberal ideology preaching supply-side policies and competitiveness. States should not protect industries unable to compete in the international market, but rather roll back its interventions in the economy. The emphasis put on exports and attracting foreign direct investments led to efforts to boost the competitiveness of localities, thus triggering a race to the bottom in labour legislation and wages (Chan and Ross 2003, Silver 2003). By dispossessing workers of their legal protection, but also of jobs through relocation, workers become, according to Harvey (2003), the equivalent of a reserve army. Trade unions can thus be held in check, paving the way for the needed rise in the rate of exploitation. This new accumulation regime has provided a common experience for workers worldwide through, for instance, attempts at disbanding unions by casting them as scapegoats for economic recession.

It is against this backdrop we understand the growing interest for labour issues in Human Geography in general, together with a growing concern with the agency and resistance of labour in the global south (Kelly 2001, Hale and Wills 2005, Cumbers et al. 2008, Lier 2007, Webster et al. 2008, Ferus-Comelo 2009). In the current economic climate of recession and faltering recovery, it is the aim of this book to develop a better and fuller appreciation of labour market processes and regulation in developed and developing countries alike. We seek to start a discussion as to how theoretical perspectives on both labour in general, and the organizations of the labour movement in particular, can be refined and redefined. Hence, a major objective is to investigate, systematize and further develop significant insights gained by geographers engaging with labour over the last couple of decades. In so doing, several of the contributions explore how theoretical developments can be made by revisiting classical texts on labour and cross-fertilizing them with recent debates in geography.

As the title of this book implies, the contributors have made an attempt to further the work started by Andrew Herod, who in two seminal articles from 1995 and 1997 introduced the concept of Labour Geography. His point of departure was a critique of both the neoclassical Economic Geography of the 1960s and 1970s, and the Marxist Economic Geography of the 1980s for a deficient attention to labour. In his view, neoclassical Economic Geography is 'fundamentally descriptive of labor, and it examines the geography of economic activity from the explicit perspective of capital' (Herod 1997: 7). Likewise, he argues that the priority given to capital in Marxist geography renders labour a reactive agent only able to resist, at best, and not to take the initiative in struggles. Herod thus points to the need for conceiving of labour as a (pro)active agent in the production of economic geographies and not merely a passive victim of the dictates of capital. Hence, labour ought to be examined as an agent that can make significant changes for its own sake, and workers are seen as capable of shaping the economic geographies of capitalism based on their self-reproduction on a daily and generational basis. Important in this regard is that the agency of labour may not necessarily be expressed only through their collective organization, a supposition which means that we also need to pay attention to the agency of unorganized labour (Kelly 2002, Rogaly 2009). Moreover, Herod (1995) argues that neglecting the agency of workers has grave implications since it is not a far cry from implying that their actions are inconsequential, an assumption which may lead to political resignation. However, Herod does agree with Marxists that there are limits to the agency of labour, although adding that the same applies to capital.

In sum, Herod argues that this warrants a new approach to labour that he refers to as Labour Geography, which contrasts the more passive approach that the conception of geography *of* labour entails and that the term depicts. However, despite allegiances to Labour Geography as a source of inspiration it varies to what extent and in what ways labour is given analytical primacy. Jonas (1996, 2009), for instance, is more concerned with the 'structures and constraints under which wage labor exists "despite itself"' than 'labor as an economic and political agent "for itself"' (Jonas 2009: 62). Similarly, being concerned with labour regulation at the macro-scale, Peck (1996) can be said to apply a political economy approach inspired by the regulation school.

Labour Geography contributes to both the discipline of Economic Geography and the field of labour studies in two important ways. First, as already seen, the emphasis on the agency and interests of labour is an important corrective to the one-sided focus in Economic Geography on understanding and representing the interests of capital. Second, Labour Geography provides an approach to place, space and scale that not only under-labours explanations of the plight of workers but also opens for contestation and analyses of strategies of resistance (Herod 1995, 1997, 2001). Hence, space is regarded as a source of power and is thus not only conceived of as an arena or reflection of society, but also as fundamental to the constitution and functioning of society, as seen for instance in Harvey's (1982) concept of 'spatial fix'. Herein lies a source of potential struggle since the geographies produced by capital attenuate the internal contradictions of capitalism, paving the way for increased profits, and may thus be contrary to the geographical visions of labour (Herod 1997). Such understandings have been a source of inspiration also for researchers in the wider realm of labour studies, especially the understandings 'of how markets, governance and social responses are embedded in place': that processes and events at the local scale should be linked with processes and events at the global scale, that labour is not only local and capital not only global, and of 'spatial fixes' as a solution to the problems of capital accumulation (Webster et al. 2008: 14).

According to Wills (2009), Labour Geography has matured over the years and has gradually widened its thematic approach to encompass most of the significant debates within the wider field of Human Geography and has thus become more mainstream. Among the themes are public policies that affect labour, work and identity, class and identity, the place of labour in multi-scalar civil society and global justice. In the same vein, Castree (2007) argues that Labour Geography as a field has few analytical boundaries. On the other hand, theoretically he characterizes the armoury of Labour Geography as heavily influenced by Marxian, feminist, anti-racist and institutionalist approaches. As a corollary, Labour Geographers are more often than not on the left side of politics with a continuing commitment to critical theory. In his view, Labour Geographers need to ask systematic questions about the content and aims of Labour Geography, and to meet analytical challenges pertaining to transnational scales of organizing, connections between trade unions and new social movements, the micro-geography of employment, how employers, as well as workers, use geography in struggles, labour migrants and to questions arising in the global south. More specifically, Castree (2007) argues that there are seven areas in which Labour Geography is still quite undeveloped: agency is still under-theorized and under-specified, migration is not given due attention as a topic of analysis, the state is weakly thematized and theorized, geographical concepts are not synthesized, wider questions of how people live and seek to live are not addressed properly, moral geographies are taken for granted and Labour Geographers need to address the normative side of Labour Geography which is weak in areas such as evaluation and policy. Lier (2007) adds to these the necessity of including attention to often-neglected groups, voices and places, such as for instance workers in the informal economy, and more research on issues pertaining to labour in the global south. Furthermore, he also criticizes the overemphasis on studies of labour in the manufacturing sector.

The aim of this book is to contribute to the on-going discussion of what issues need to be dealt with in future research. The emphasis of the book is on labour, which is theorized as an active agent able to mobilize and become a social force resisting the strategies of states and companies. However, we argue that this requires a more systematic theorization of the agencies of the state and capital as well, since agency is understood as relational. In relation to the above calls made by Castree (2007) and Lier (2007), we want to engage in the debate on agency, discussing the collective and individual agency of labour.

Agency is thus a recurrent theme in this volume and we would like to provide some introductory remarks on our standpoint. In line with Jary and Jary (2000: 9), we understand agency as human action that makes a difference to social outcomes, and as 'the power of actors to operate independently of the determining constraints of social structure'. In other words, social structures do not determine how agents must think and act, although they do define certain situational logics (Creaven 2000). Hence, agents and agency are not structurally determined even though they are structurally conditioned. Furthermore, agents may exert influence on the existing structures either through reproducing or changing them. The interplay between agents and structures can be understood as a never-ending series of cycles consisting of structural conditions, social interaction and structural development.

Furthermore, we have with this book contributed to an expansion of Labour Geography's study area, geographically to the global south, and analytically to encompass the challenges of organizing workers in the informal economy as well as temp workers. Agency might thus be linked to geography, and North America and Europe have for quite some time represented the empirical 'heartland' of research in Labour Geography. While the agency of workers have, as we have seen, been emphasized, this pertains largely to workers in the global north, while early Labour Geographers tended to regard workers in the global south as passive victims of imperialist capital (Herod 1997). Moreover, our case studies comprise both the manufacturing and the service sectors. Some of the contributions to this book also demonstrate the merit of the concept of labour regimes, primarily the analytical value when explaining working conditions, but also to some extent regarding the lives of workers more holistically.

As Labour Geographers, we discuss the concepts of space and spatial fix in relation to the agency of labour and address issues of union renewal. We agree with Herod (2003) that scale should not solely, or even primarily, be conceived of as a hierarchical ladder ranging from the local to the global, but rather as a network consisting of horizontal relations as well as vertical. Consequently, horizontal relations between workers and their organizations in two or more places represent a form of upscaling and may be considered a starting point of a global response despite being based on local-local linkages (Webster et al. 2008). This becomes all the more important if we heed Webster et al.'s (2008) argument that capital needs to be confronted at multiple scales simultaneously, and that it is not necessary for labour to conquer local, regional, national and global scales in a sequential manner.

The book derives from the conference *Theoretical Approaches in Labour Geographies* arranged by the Department of Sociology and Human Geography, University of Oslo 14–15 May 2008 in collaboration with the IGU Commission on the Dynamics of Economic Spaces. However, in a fast-growing field of research, a single volume of seventeen thematic chapters cannot provide an exhaustive account or offer full justice to all of the above issues. Hence, keeping to the analogy of missing links, we have attempted to assume the role of a goldsmith linking contributions that in different ways highlight the various issues, complement each other at the thematic level and at the same time explicitly or implicitly link to other contributions across the thematic divide rather than a welder seeking to produce a solid and uniform cast. This breadth of approach serves to underscore the complexities of the issues which Labour Geographers deal with. The group of authors comprises internationally renowned researchers in Labour Geography, established researchers with an interest in labour research and young researchers.

Plan of the Book

Part I of this book, 'Linking Approaches in Labour Geography', sets the agenda by discussing where Labour Geography stands today and by identifying its challenges with an emphasis on the agency and spatiality of labour.

In Chapter 2, Herod maps the genealogy of Labour Geography and points to some of its lacunas. Concurring with Castree (2007), Herod argues that despite the emphasis on workers' agency being a main contribution, it is also one of the areas which requires attention in the future. Labour Geography started out as an attempt to qualify the crude argument that while capital is capable of commanding space, workers are passively trapped in place. First, Herod argues that neither of the two agents are monoliths. Different segments of capital have 'different degrees of place fixedness', while some workers are more geographically mobile than others. Such differences also mean that workers have to negotiate spatial interest and class interest. Second, he re-emphasizes his earlier argument that workers are active geographic agents striving to create their own spatial fixes, although they cannot do so under conditions of their own choosing. Consequently, space is not simply a container or an expression of social processes, but is actively produced, i.e. there is a politics to space (above). In Herod's view, agency is usually employed in a not very nuanced manner, and he proposes to differentiate it along Aristotelian lines. Aristotle differentiated causality into material, formal, efficient and final. This means that causal powers derive from components (material cause), conceptions, triggering events, and overall purpose (telos) respectively. Furthermore, Herod challenges fellow Labour Geographers to expand their empirical basis geographically and historically to encompass other social formations than those of the industrial capitalist societies and socially to include non-work spaces and non-unionized workers.

In Chapter 3, Coe and Jordhus-Lier search for an understanding of the agency of labour in the context of neoliberalization. Their major argument is that the contemporary fragmentation of labour agency as a political project necessitates theoretical antidotes to the conceptual fragmentation of labour agency. They recommend Labour Geographers not to let go of the insights established in Labour Geography over time; that all workers are embedded in structures of capital, state and community, and that labour agency is a relational concept. Unpacking agency along the dimensions of geography, capital is discussed in terms of global production networks and the state is analysed in terms of its different roles as regulator, container of labour practices, provider of basic services, political apparatus, employer, and boss. In their discussion of community, they advocate a holistic understanding of the worker. Workers are embedded in local communities, which they rely upon when the workplace is put at risk. Communities also represent opportunities for mobilization along many lines. They are sites of recruitment and organization around new scales where identification of alternative targets for action is undertaken. The approach advocated by Coe and Jordhus-Lier serves to highlight that workers' room for manoeuvre, and thus labour agency, varies immensely in time and space, and that our understanding of it necessitates multiple theoretical tools.

Early Labour Geography did not differentiate between workers as individual agents and workers as collective agents. Differentiation between union leadership and its rank-and-file membership also came later. Hence, Part II of this volume focuses on the collective agency of labour and indicates a growing interest in research on the relevance of traditional unions and their changing strategies in the present national and global economic contexts.

In Chapter 4, Andrew Cumbers and Paul Routledge explore how trade unions attempt at meeting the challenges of globalization of capital and new networks of production by forming global union federations (GUFs). They introduce the concept of 'entangled geographies' to denote the complexities revealed when strategies and practices of GUFs are studied. The authors find that global connections of unions open up possibilities for trans-national labour solidarity, but that different subject positions of place-based, though not bounded, social actors infuse union operations. They show how broad visions of solidarity collide with local and national interests. This reflects the spatial and organizational logics of the union movement, the weaknesses of top-down models of solidarity where national and international levels dominate. The authors therefore advocate more locally based and grassroot forms of union action.

In Chapter 5, Ryland delves into the issue of international solidarity by examining how trade union agendas that aim to promote labour internationalism are perceived by union members. She argues that Labour Geography has failed to look beyond workers as economic beings and that it has not really managed to grasp the agency of workers. With the theoretical point of departure that workers should be studied as social beings who possess a number of different identities, she concludes that union membership is not automatically conducive to altruistic solidarity with global partners. Her findings corroborate Hyman (1999) that labour solidarity requires construction by political communication and education. While executives in the union that she studied supported internationalism, grassroots members were more sceptical and perceived it as a top-down process. However, when commonalities with workers abroad were identified in discussions on internationalism, grassroots members also became more supportive. This way she has also identified different collective agencies within one union and reasons why it occurs.

Theoretically Chapter 5 draws on the notion that workers are social beings with multiple identities. In Chapter 6, Bergene elaborates this in the discussion of how to navigate the chaotic consciousness of the trade union movement. Bergene revisits the long-standing debate on labour consciousness and she argues that much of the debate is marked by an overly structural determinism. By letting the philosophy of critical realism underlabour a revised framework, these theories are refined by the introduction of the term *chaotic consciousness*. The chapter argues that what has been termed trade union consciousness is not really a particular consciousness as such, but rather an expression of a chaotic consciousness derived from structural positionings, ideology and other concrete, and in some instances contingent, factors pertaining to the individuals working in the union bureaucracy. Finally, the author provides some reflections on how unions may navigate this chaotic consciousness, drawing on Gramsci's discussion of democratic centralism in political parties and Freire's Dialogical Method. Chapter 6 thus cross-fertilizes theories on trade unions with perspectives from both studies of political parties and pedagogy.

Chapters 7, 8, and 9 deal with different aspects of trade union renewal at the national scale. This ranges from new ways of organizing and campaigning that are characteristic of Schumpeterian unionism, to recruitment of temporary workers and merger between organizations in the formal and informal economy.

In Chapter 7, Tufts challenges the binary conception of business unions and social movement unions based on a study of the Canadian branch of UNITE-HERE. This is a union formed through a merger of textile (UNITE) and hotel (HERE) workers in 2004. By careful comparison between the specific union practices of UNITEHERE and the ideal-type of defensive Atlantic unionism and the ideal-type social movement unionism, he has developed an analytical framework to examine and explain labour union renewal, what strategies are applied and the effectiveness of these to hold the line against neoliberalism. Among the parameters looked into are intra-institutional organizing in terms of recruitment and servicing and collective bargaining, extra-institutional organizing with emphasis on coalition building, international solidarity and mergers, and what are the central labour bodies in this. In addition, the analytical framework includes labour-management relations, such as cooperation and training, and labour-state relations in the areas of economic development and labour market regulation. The UNITEHERE is then conceptualized as a Schumpeterian union. Tuft's work on Schumpeterian unions continues Jessop's (1993) theorizing of the Schumpeterian workfare state.

Contemporary trade unions face a dual challenge of reduced membership and increase in the number of casual workers that is difficult to organize at industrial sites. They need to develop strategies to kill these two birds with one stone, i.e. organizing temporary workers and thus get new members. In Chapter 8, Meyer and Fuchs apply the concept of dynamic capabilities, which was originally developed to understand learning processes within large organizations, to understand this learning process. The concept guides the investigation of how unions adapt to changes in socio-economic framework conditions. Meyer and Fuchs exemplify how dynamic capabilities may develop, how they vary regionally and at multiple scales, and discuss the relative merits of structural, organizational and person-oriented perspectives in explaining their findings. Their contribution demonstrates the merits of being theoretically eclectic; searching for fruitful perspectives within organizational sciences and Economic Geography.

Another attempt at trade union renewal is addressed by Andrae and Beckman in Chapter 9. The recent efforts of the Nigerian textile workers union (NUTGTWN) to negotiate a merger with an association of local tailors have inspired the authors to discuss the potential of organizing across the formal-informal divide. With the industrial collapse in Nigeria unions are losing members and it is a big challenge to protect the rights and welfare provisions that the workers have attained. At the same time, the informal economy suffers difficult working conditions and limited social welfare. Social policy is important to the welfare of both workers and the unemployed as well as the credibility of unions. Hence, the authors argue that a merger with associations in the informal economy could focus on welfare services such as water, health and education. These are arenas in which the state has proved incompetent. The function of the merger will thus be to voice demand and to discipline state institutions through pressure from below. The contribution is hypothetical in its approach, but contributes to a most important, but unfortunately rather under-researched subject area, in Labour Geography.

In Part III on the politics of labour, relations between agencies of collective labour, agencies of the state and agencies of capital are played out. Chapters 10 and 11 pay particular attention to contextual conditions and draw on Gramsci's notion of an historic bloc in this endeavour. While contextual conditions are important in Chapter 12 and 13 as well, the main focus is on labour regimes. While Chapter 12 highlights the importance of paying attention to welfare regimes and political regimes in explanations of changing labour regimes, Chapter 13 discusses links between the national scale and the local scales in labour regulation and how this affects local labour control regimes. It also discusses how and why different types of capital influence local labour control regimes at the workplace level.

More concretely, in Chapter 10 Jauch and Bergene provide an analysis of the contextual embeddedness of trade unions and how the involvement of trade unions in struggles for independence, or national-popular struggles as Gramsci terms them, impacts upon their subsequent development. Drawing on lessons from Namibia, the authors argue that trade union politics and strategies must be analysed in relation to the socio-political context since trade unions do not make independent decisions formulated in a vacuum. Merely emphasizing the agency of workers and unions is thus not sufficient, since the way this agency is exerted depends to a large extent on contextual conditioning.

Taking Harvey's concept of spatial fix as a point of departure, Doucette argues in Chapter 11 that attempts at understanding the agency of labour need to delve more deeply into the political processes external, though related, to capital accumulation. In order to undertake this task, Doucette proposes a framework combining the insights gained through theories within the field of political economy on spatial fixes with a Gramscian perspective on hegemony and civil society. Most importantly, Doucette calls for abandoning abstract analyses of class relations in which capital and labour are reduced to economic agents in favour of analyses of the multiple sets of power relations in concrete social formations. As such, the agency of labour cannot be regarded as confined to the tension between wages and profit, but rather rendering labour both an economic and a political agent. In Doucette's view, Herod's attempt to further develop Harvey's concept of spatial fix to include the agency of labour has strengthened our understanding of the relation between capital accumulation and the wider socio-political expanse in which it is embedded. His analytical framework is employed in a case study of labour relations in South Korea, which addresses the transition from an authoritarian regime exerting labour control through a coercive apparatus to a more democratic regime. For instance, by establishing bodies for social cooperation, civil society organizations such as unions are directed by the state through consent.

In Chapter 12, Knutsen and Hansson explain labour activism in transition economies with reference to both the micro-politics of the workplace and the macro-politics of the state, commonly referred to as labour regimes. It is argued that in studying the power relations between state, capital and labour, their spatial embeddedness must be addressed. Providing empirical examples from the transition economies China and Vietnam, the chapter argues that processes of commodification and decommodification of labour and welfare are of particular importance when trying to understand the dynamics of labour regimes and labour activism. Studying transition economies provides the authors with a unique vantage point, since the negotiation between commodification of labour after state socialism and processes of decommodification through regulation are of pivotal importance for labour's vulnerability and thus leverage. In the same vein, it demonstrates the complexities of power relations between labour, capital and the state.

In Chapter 13 Magnusson, Knutsen and Endresen address how and why different workplace regimes develop in the same industrial sector in one local society. The chapter demonstrates how Gramsci's theory of hegemony and historic bloc can be applied as a theoretical tool in analysing national labour relations and its links to the local scale, and to analyse relationships between capital and labour at the workplace. However, in order to explain differences in local workplace regimes it is necessary to pay attention to the competitive strategies of the firms in question and the nature of their embeddedness in local society. This way the chapter also contributes to theorization of the agency of capital.

A political economy of labour approach binds the three contributions of Part IV on labour and the strategies of capital together. At the same time, one may argue that all of the three chapters address the need of unpacking capital and discussing it in relation to the agency of labour, although to a varying degree. Based on cases in the global north and global south respectively, Chapters 14 and 15 address how international competition and industrial restructuring affect the relationship between capital and labour. Chapter 16 adds directly to the agency debate by exploring how theories of alienation explain the relationship between capital and labour and the increasing reliance on labour hire companies for employment.

In Chapter 14 Taylor and Bryson explain the phenomenon of 'erosion from below' in the manufacturing industry of the UK. Their point of departure is that the UK management and the UK government since the Thatcher period have been selling off national assets. This has pushed UK engineering firms down the value chain where prices are squeezed. Hence, training of labour is left to the government whose training programs are poor and disconnected from the real world of work. The result is dependence on old workers and difficulties in recruiting young workers with necessary competence and skills. From an agency point of few, this case demonstrates how the strategies of capital and government together result in a price squeeze that not only affect labour negatively, but also affect capital and local economic development negatively in the longer run by eroding the sustainability of local manufacturing.

While early segmentation theories were preoccupied with processes at the national scale, Beerepoot, in Chapter 15, addresses the outcome of contemporary globalization on segmentation of labour in the Philippines. This is a context in which workers experience limited formal protection and representation, and they are highly vulnerable to international shifts in production. Beerepoot draws on Peck (1996) that processes of segmentation take place at different geographical scales and that these have different outcomes in particular sites. Thus, workers are at the same time part of labour markets that are segmented at the international, national and local level. The new international division of labour (NIDL)-thesis and newer literature on the second global shift inform critical questions regarding what types of jobs are vulnerable to international outsourcing and consequences of this to workers in the countries that the jobs are shifted to. His empirical work focuses on the decline of the import-intensive export sector and the concomitant growth of jobs resulting from off-shoring of service sector activities, more precisely the sub-sector of call centres. A new labour segment is created nationally for young people who may earn a high income, but the bulk of the jobs are low-skill customer-care jobs that are uncertain in a longer term perspective. Hence, the author argues that workers who lose their job due to the decline in manufacturing industry have limited access to new jobs and resort to the informal economy.

In Chapter 16 Endresen theorizes labour hire agencies. Taking a remark made by an informant as a point of departure, she explains why labour hire agencies exist and thrive in contemporary capitalism, what perspectives on social reality they necessitate, and how the phenomenon affects social existence. By employing classical theories on reification and alienation, Endresen argues that the increased use of labour hire agencies bears witness to a purification of the commodity form of labour, a process which rests on reification and furthers alienation among capitalists and workers alike. Under such conditions, labour power is treated as a commodity like any other. Its sentient nature, and thus its pseudo-commodity character, is suppressed by the understanding of labour power as embedded in bodies irrelevant to the task at hand. By digging deeply into such abstract theories, Endresen helps develop a wholly new perspective on the phenomenon of labour hire.

In Chapter 17 which is the concluding chapter of the book we discuss what we think is still missing in Labour Geography. We relate this to the above suggestions made by Castree (2007) and Lier (2007) and Coe and Jordhus-Lier in Chapter 3, as well as the lacunas in Labour Geography that Herod identifies in Chapter 2.

Acknowledgements

First of all, we would like to thank Professor Richard Le Heron and Professor Mike Taylor of the IGU Commission of the Dynamics of Economic Spaces for constant encouragement and support from the conception of the idea of arranging a meeting on Labour Geography to the completion of this volume. In addition, Mike Taylor has helped us with the language for which we are most grateful. We would also like to thank the international reviewers for their useful comments on the individual chapters, Hallvard Berge for his technical and secretarial contribution, and Annika Wetlesen at the Department of Sociology and Human Geography for inspiring discussions and valuable comments whenever we needed feedback and a second opinion. Last but not least, we are thankful to the Department of Sociology and Human Geography, University of Oslo, for practical and financial support.

References

Castree, N. 2007. Labour geography: A work in progress. *International Journal of Urban and Regional Research*, 31(4), 853–62.

Chan, A. and Ross, R.J.S. 2003. Racing to the bottom: international trade without a social clause. *Third World Quarterly*, 24(6), 1011–28.

Creaven, S. 2000. *Marxism and Realism. A Materialistic Application of Realism in the Social Sciences*. London: Routledge.

Cumbers, A., Nativel, C. and Routledge, P. 2008. Labour agency and union positionalities in global production networks. *Journal of Economic Geography*, 8, 369–87.

Ferus-Comelo, A. 2009. Mission impossible? Raising labor standards in the ICT sector. *Labor Studies Journal,* 33(2), 141–62.

Hale, A. and Wills, J. (eds), 2005. *Threads of Labour*. Oxford: Blackwell.

Harvey, D. 2000. *Spaces of Hope*. Edinburgh: Edinburgh University Press.

Harvey, D. 2002. *The Limits to Capital*. Oxford: Basil Blackwell.

Harvey, D. 2003. *The New Imperialism*. Oxford: Oxford University Press.

Herod, H. 1995. The practice of international labor solidarity and the geography of the global economy. *Economic Geography*, 71(4), 341–63.

Herod, A. 1997. From a geography of labor to a labor geography: labor's spatial fix and the geography of capitalism. *Antipode,* 29(1), 1–31.
Herod, A. 2001. *Labor Geographies. Workers and Landscapes of Capitalism*. New York: Guilford Press.
Herod, A. 2003. Geographies of labour internationalism. *Social Science History,* 27(4), 501–23.
Hyman, R. 1999. Imagined solidarities: Can trade unions resist globalization?, in *Globalization and Labour Relations*, edited by P. Leisink. Cheltenham: Edward Elgar.
Jary, D. and Jary, J. 2000. *Collins Dictionary of Sociology*. Third edition. Glasgow: Harper Collins Publishers.
Jonas, A.E.G. 1996. Local labour control regimes: uneven development and the social regulation of production. *Regional Sudies,* 30(4), 323–38.
Jonas, A.E.G. 2009. Labor control regime, in *International Encyclopedia of Human Geography*, edited by R. Kitchin and N. Thrift. Amsterdam: Elsevier, 1–7.
Kelly, P.F. 2001. The political economy of local labour control in the Philippines. *Economic Geography*, 77(1), 1–22.
Kelly, P.F. 2002. Spaces of labour control: comparative perspectives from Southeast Asia. *Transactions of British Geographers,* 27, 395–411.
Lier, D. 2007. Places of work, scales of organising: a review of Labour Geography. *Geography Compass*, 1(4), 814–33.
Lipietz, A. 1982. Towards global Fordism? *New Left Review,* 32, 33–47.
Peck, J. 1996. *Work-Place. The social regulation of labor markets*. New York: Guilford Press.
Rogaly, B. 2009. Space and everyday life: labour geography and the agency of unorganized and temporary migrant workers. *Geography Compass*, 4(3), 1–13.
Silver, B.J. 2003. *Forces of Labour. Workers' Movements and Globalization since 1870*. Cambridge: Cambridge University Press.
Webster, E., Lambert, R. and Bezuidenhout, R. 2008. *Grounding Globalization. Labour in the Age of Insecurity*. Oxford: Blackwell Publishing.
Wills, J. 2009. Labour Geography, in *The Dictionary of Human Geography*, edited by D. Gregory, R. Johnston, G. Pratt, M.J. Watts, and S. Whatmore. Chichester: Wiley-Blackwell, 404.

Chapter 2
Labour Geography: Where Have We Been? Where Should We Go?

Andrew Herod

Introduction

I initially pondered whether there was anything to say about Labour Geography, for in the view of some commentators both labour and geography seem to have become irrelevant of late. On the one hand, we have heard in recent years that the organized labour movement is dying, if not dead, and that workers, at least within neoliberal eyes, should not be seen as labourers (far too much of an antediluvian identity for such times of alleged market triumph!) but, instead, as consumers and/or members of employment teams, teamwork being the leitmotif of contemporary discourses about work. On the other hand, in the age of globalization, geography, it has been argued by some, has been annihilated; hence the litany of books heralding the 'death of distance' and the emergence of a 'flat and borderless' world (Cairncross 2001, Friedman 2005, Ohmae 1990). Thus, it is contended, the revolutions in transportation and telecommunications technology that are transforming the spatial and temporal organization of capitalism mean that '[i]n the age of the global economy, physical location is much less important [than previously and it] no longer matters where a company is based' (Ohmae 2005: 13, 94). Within this situation, few places are now more than 24 or 48 hours apart, such that São Paulo or Guangzhou can be considered manufacturing suburbs of Los Angeles, London, Sydney, or Paris. In such a world, neoliberal globalization discourse suggests, capital is active and can overcome space, the erasure of geographical boundaries is automatically leading to greater competition for many workers, who are passive and confined to place, and the best thing that workers can do is to get on the neoliberal train or be left at the station.

However, a quick look around the world reveals a different picture. Workers continue to organize against the depredations of global corporations, whether it is Korean workers challenging the power of the Chaebols, Argentinian workers taking over abandoned factories, or Chinese workers beginning to agitate for their rights in the vast factory zones of southern China. Equally, the shrinkage of relative distances between places and the speed-up of contemporary economic and political life, a process which Harvey (1989) has referred to as 'time-space compression', is a highly geographically and historically uneven process. It is a process, in other words, in which geography and spatial differentiation are central.

Moreover, it is a highly contradictory process, for in many cases, as spatial barriers diminish, geography actually becomes more important, not less. When capital can locate anywhere globally, where it actually does choose to locate is often on the basis of small differences between places. Additionally, it is a process which both dramatically affects workers and in which workers themselves are centrally involved. Thus, not only does the shrinking of the globe mean that local workers may indeed increasingly have to compete with those located in communities on the other side of the planet, but it also can open up opportunities for them actually to come into greater contact with their fellow workers in such communities, perhaps making it easier for them to develop trans-spatial solidarities. At the same time, given that worker solidarity efforts are all about 'coming together [and] organizing over space' (Southall 1988: 466), workers have themselves been intimately involved in bringing about some of the transformed geographical relations between places which are at the heart of the phenomenon of the shrinking globe (see Herod [1997a] for a detailed case study of how the US labour movement helped US capital expand into Latin America and the Caribbean, thereby integrating more closely the US economy with this region). These developments suggest, then, that something geographical is clearly going on with regard to contemporary transformations in the nature of capitalism, and that workers are deeply involved in it. Equally, whereas geography has often been seen merely as a context or a stage upon which social life plays out, in fact the making of the geography of capitalism is a highly political act. Consequently, as contemporary processes restructure the time-space organization of global capitalism, being spatially aware will be an important element in workers' political and economic praxis, at scales from the very local to the truly global.

Given that neither labour nor geography seem to be disappearing, then, in this chapter I want to assess the current state of Labour Geography. The chapter is organized as follows. First, I detail how Labour Geography emerged as a field within Anglo-American geography and how it has been taken up in parts of the non-English speaking world, as evidenced by its growing popularity in countries like Norway and Germany – witness the holding of the conference at the University of Oslo which served as the catalyst for this book and the advance of the field of *Geographie der Arbeit und Beschäftigung* (geography of work and employment) in Germany (see Belina and Michel [2007] and the 2002 essay by Berndt and Fuchs). Second, I outline some of Labour Geography's central axioms. Finally, I ponder some of Labour Geography's lacunae.

Labour Geography's Genealogy

Arguably, the central tenet in Labour Geography is the claim that the geography of capitalism makes a difference to workers and that workers make a difference to the geography of capitalism. To understand how this central tenet came about, one must understand the field's genealogy, for Labour Geography's current theorization of both workers and the making of capitalism's geography is a reaction to how both labour and economic landscapes had long been thought of within Economic Geography. In particular, it is probably fair to say that Labour Geography has its ultimate origin in a reaction to the type of Economic Geography practised when neo-classical economics and the spatial science tradition dominated the field, roughly the period from the mid-1950s to the mid-1970s. Specifically, within the neo-classical, spatial science tradition labour was frequently thought of as being no different from any other factor of production like the cost of land, raw materials, or energy, a commodity to be simply bought and sold (see Barnes [2000] for more on Anglo-American Economic Geography's disciplinary history). In such a view, labour was incorporated into explanations of how economic geographies are made from the point of view of corporate decision-makers seeking to locate plants and so workers were considered simply in terms of the geographic variation of their cost, their degree of political organization, their skill level, and so forth. As Massey (1973: 34) put it, in such an approach 'profit is the criterion, wages are simply labour costs.' Likewise, landscape was thought of as little more than a sterile stage upon which economic relationships are played out according to various laws which could be described in mathematical terms. In such a view, social life did not produce space but instead merely rearranged objects within it. This was, then, the classic Newtonian approach to space, one which sees it as a container within which social objects exist but from which they are separate (for more on various conceptions of space, see Curry [1996]).

In reality, of course, labour *is* quite different from other factors of production and commodities, if for no other reason that it is sentient and can thus shape the accumulation process, a fact which led Storper and Walker (1983: 4) to declare labour a pseudo-commodity, one which is 'idiosyncratic and spatially differentiated'. Likewise, the way in which the economic landscape is configured can have great political and economic import. Hence, employers may locate work in isolated places so as to prevent workers from coming into contact with one another or may seek out new, non-union locations in which to locate new investment. The founding of myriad US suburban manufacturing towns in the late nineteenth/early twentieth century, for example, was the result of companies seeking to escape urban unions (Gordon 1978). Consequently, beginning in the late 1970s Marxist geographers began to examine more closely the making of the geography of capitalism and the place of labour within that. For instance, Massey (1984) argued that capitalist accumulation spawned particular spatial divisions of labour at particular times and that these shaped subsequent patterns of investment. Hence, the north of England was industrialized in the nineteenth century, deindustrialized in the early post-WWII period, and then the large pools of unemployed labour that deindustrialization created proved attractive for subsequent investors looking for locations for light manufacturing. For his part, Harvey (1982: 233) suggested that capital needs a certain *spatial fix* of investment to ensure that accumulation can occur. Raw materials and workers need to be brought together in particular locations, a fact which spawns the construction of various types of infrastructure, like the 'factories, dams, offices, shops, warehouses, roads, railways, docks, power stations, water supply and sewage disposal systems, schools, hospitals, parks, cinemas, restaurants – the list is endless' that enable the securing and realization of surplus value. Finally, Smith (1984) outlined how internal contradictions within the structure of capital, particularly its need to be fixed in space so that accumulation can occur but also its desire to remain mobile so as to take advantage of opportunities arising elsewhere, was the driver of the uneven geographical development that is the hallmark of the economic landscape under capitalism.

Nevertheless, despite conceptualizing labour in more complex terms than did the spatial scientists whom they were challenging, as an idiosyncratic and spatially differentiated pseudo-commodity *à la* Storper and Walker, for instance, Marxist geographers generally similarly viewed the making of the geography of capitalism from the point of view of capital. Labour was still seen as little more than '*variable capital*, an aspect of capital itself' (Harvey 1982: 380–81, original emphasis) and workers still largely entered 'the theater of history as abstract labour, factors of production, dependent variables in the grand narratives of crisis and renewal' (Aronowitz 1990: 171).[1] This was, then, still a Geography *of* Labour (Herod 1997b) and although worker organization could be seen in a reactive sense to play an important role in how capitalism's landscapes evolved – Peet (1983), for instance, outlined how capital avoided areas of strong labour organization – it was capital which was viewed as the only proactive agent shaping economic landscapes' form. This is, perhaps, hardly surprising, given that Marx himself examined capitalist accumulation from the viewpoint of capitalists. However, the result was that the geography of capitalism was understood to be a reflection of the actions of capitalists – as Harvey (1978: 124) put it, it was capital that 'represents itself in the form of a physical landscape created in its own image [and which] builds a physical landscape appropriate to its own condition at a particular moment in time.' Thus, how the geography of capitalism is produced was increasingly conceptualized as an integral part of the accumulation process (i.e. capitalists must produce landscapes of profitability rather than unprofitability), but workers were viewed as generally confined to place whereas capitalists were viewed as commanding space, an image that has played into discourses of globalization that rely upon the notion that capitalists can overcome space/distance but that workers are place-bound.

By the late 1980s/early 1990s, though, a number of scholars were beginning to question such understandings and dualisms. Thus, the contention that capital is *de facto* more mobile than workers was challenged as too simplistic. Indeed, as Marx (1858 [1973]: 548–549) himself had argued, within the accumulation process capital is fixed in time at least part of the time ('it must spend some time

1 Although Aronowitz was talking about how many Marxist historians had conceptualized labour in their writing about the history of capitalism, much the same could have been said of Marxist geographers writing about capitalism's geography.

as a cocoon before it can take off as a butterfly') and this usually implicates place in some fashion, given that accumulation does not take place on the head of a pin. Equally, capital needs to *work place* if it is to be successful: it must work to embed itself locally so as to develop the economic relationships with local labour forces, elected officials, suppliers, and so forth necessary to ensure that accumulation and the realization of surplus value can occur, a practice which may take considerable time. Some segments of capital, e.g. utility companies or steel mills, are also highly place-dependent because of the size of their sunk costs, a fact which can lead them to engage in place-based boosterist politics because of their local dependence and inability to flee a particular location should the local economy sour (Cox and Mair 1988). Similarly, workers are not necessarily unable to command and/or cross space, as the history of labour migration and successes in developing trans-spatial, and often international, solidarity efforts illustrate. The idea that capital can overcome space but that labour is fixed in place, then, came to be seen as too crude. Rather, different segments of capital and labour have different degrees of place fixedness and ability to cross space – some workers are more geographically mobile than are some types of capital – with the result that there is obviously a politics to place and space and the making of economic landscapes. Perhaps the most significant development, though, was the recognition that not only do capitalists need to ensure that the economic landscape is structured in particular ways so that capital accumulation can occur (in other words, so that capital can reproduce itself), but that workers, too, need to ensure that the landscape is made in such a way that they can secure their own social and biological reproduction on a daily and generational basis. This requires that particular spatial configurations of what they need to so do – jobs, homes, shops, schools, recreation facilities – are emplaced in the landscape.

Acceptance of the argument that shaping the landscape in particular ways is key to labour's self-reproduction had several consequences. First, it presented workers in a new theoretical light, that of active geographical agents. Such a conceptual shift has been characterized as marking a transition from viewing workers within the framework of a Geography of Labour to doing so through that of Labour Geography. Rather than viewing labour from the perspective of how capitalists choose between different groups of workers when making locational decisions, then, Labour Geography has focused instead upon how workers develop their own spatial fixes, how they attempt to implement spatial strategies as part of their political economic struggles, and how they thereby seek to shape the economic geography of capitalism to their own ends, all the while recognizing that, paraphrasing Marx, whilst workers may make their own geographies they do not do so under conditions of their own choosing. Second, it reinforced the idea that because workers may need a different *spatial fix* to ensure their daily and generational self-reproduction than that preferred by either capital or the state, workers', capitalists', and the state's ideal fixes may be in conflict with each other. As a result, the economic geography of capitalism is actively struggled over as part of each social actor's spatial praxis. Third, and related, rather than thinking of labour, capital, and the state as monolithic socio-spatial actors, it is obvious that different segments within each of these entities may prefer different spatial fixes. Hence, some capitalists may prefer that new jobs come to their community whilst others may not. In remote single-industry towns, for instance, new jobs may mean additional customers for retailers but may also mean that manufacturing employers lose their tight hold over the local labour market. Likewise, different groups of workers may prefer that work is located in some places rather than others, as may different segments of the state (see Herod [1998] for an example of two unions in conflict with one another over cargo handling work in US port hinterlands, and Goodman [1979] on how different state and local governments in the US have fought each other to secure investment within their territorial boundaries, thereby shaping the country's economic geography).

Workers and Space

Although it is always dangerous to try to distill a wide-ranging and diverse field's empirical studies into a small number of topics for purposes of outlining its structure, it is fair to say, I think, that five interconnected elements have dominated Labour Geography's research agenda to date.

First, there is the issue of how workers' spatial embeddedness and/or entrapment shapes their social praxis. Specifically, some workers are quite fixed in particular places due to kinship ties, their own sunk costs, e.g. homes they cannot sell, their particular skills mix which may only be useful in certain localities, and so forth.[2] They exhibit the same kinds of local dependence (Cox and Mair 1988) experienced by various segments of capital. The result is that they may engage in the kinds of boosterist local politics designed to bring investment to their communities as do local immobile capitalists and the local state, but as willing partners rather than as cultural dupes. Such actions represent a defense of spatial interests rather than class ones (Hudson and Sadler [1986] detail how understanding workers' sense of place is central to understanding reactions to European steel mill closures in 1980s).

Second, there has been exploration of how workers engage with the unevenly developed geography of capitalism. Part of this has been through examining workers' practices of developing labour solidarity, given that developing solidarity is, at least in part, about overcoming geographical distances between workers located in different communities. Thus, Labour Geographers have been quite interested in understanding how workers develop bargaining strategies which must incorporate many different sets of conditions and practices across any given

2 Southall (1988) notes in this regard that the disparate geology of coal seams in different parts of Britain led miners to develop highly place-specific skills, with the result that miners from one region could not easily move elsewhere. This contributed to highly localized work and union practices.

economic landscape. When seeking to negotiate national contracts, for instance, do union negotiators endeavor to bring all workers up to the level of the highest paid and best protected in the industry, do they try to set a minimum below which no workers should fall, or do they develop some kind of spatial average of such conditions? This raises questions about why workers choose particular spatial strategies and what consequences such strategies have for how the economic landscape evolves. At the same time, Labour Geographers have been interested in exploring how capital can use space to fetishize commodities, such as by relocating production to distant places, effectively hiding production in the global economic landscape, and how, in turn, workers seek to defetishize commodities by locating their production within, and movement across, the landscape. For instance, one area of research has concerned how workers and their organizations have mapped production chains so that they can trace commodity flows from place to place and thereby open up the economic landscape and make it more transparent. Such research has argued that explaining workers' actions frequently involves comprehending the spatial choices they face and make within the context of capitalism's unevenly developed geography.

Third, there has been a focus upon how workers make new scales of both their own social organization and that of their employers. For example, when workers seek to rework bargaining from a local to a national or even international system of bargaining, they are essentially developing a new geographical scale of organization. Hence, when US dockers shifted from a system of port-by-port bargaining to a national contract in the 1970s, they not only reconstructed their own scale of organization but forced employers to adjust theirs (Herod 1997c). By so doing, they prevented their employers from playing different ports against each other in a race to the bottom with regard to wages. However, such a national contract also established a situation in which southern dockers, who had to compete for work with cheaper, non-union workers, were increasingly priced out of the market. What this suggests is that the spatial strategies which may be successful in one era may not be so in another, and that workers must constantly readjust them as the geographical challenges they face change.

Fourth, Labour Geographers have been interested in what the spatial context within which they find themselves has meant for issues of worker identity. Hence, Mohammad (2010) has explored how Muslim Pakistani women take on different identities when they are in the spaces of the home versus those of the paid workplace, and how they negotiate their identities according to their spatial circumstances. Likewise, Stenning (2010) has questioned how the changing nature of the spatial relationships between work and community in post-socialist Poland has reshaped practices of identity formation amongst working people. Indeed, as Hyman (2004: 21–22) has suggested, changes in capitalism's spatial organization can have important consequences for worker identity. Thus, 'the spatial location and social organisation of work, residence, consumption and sociability have become highly differentiated,' such that the average employee today 'may live a considerable distance from fellow-workers, possess a largely "privatised" domestic life or a circle of friends unconnected with work, and pursue cultural or recreational interests quite different from those of other employees in the same workplace.' This spatial 'disjuncture between work and community (or indeed the destruction of community in much of its traditional meaning) entails the loss of many of the localised networks which [previously] strengthened the supports of union membership (and in some cases made the local union almost a "total institution")'. The result is that whereas formerly many workers' labouring identities 'were reinforced by the broader networks of everyday life…the possibility and character of collectivism are today very different when work and everyday life are increasingly [spatially] differentiated'.

Finally, Labour Geographers have been interested in how the changing spatialities of capitalism are encouraging workers and their organizations to develop new organizing models. For example, Savage (1998) illustrated how the Justice for Janitors campaign in Los Angeles shifted tactics from seeking to organize janitors on a building-by-building basis to organizing across entire local labour markets. For their part, Johns and Vural (2000) detailed how the UNITE garment workers union in the US teamed with the National Consumers League to confront the issue of sweatshops not by trying to organize the spaces of production directly, i.e. the sweatshops themselves, but by focusing upon the spaces of consumption and having consumers pressure large retailers not to contract with manufacturers who do not meet certain wage and workplace health and safety standards. In this context, Labour Geographers have also been interested in the rise of so-called *community unionism*, which seeks to break down the barriers between workplace organizing and organizing in the broader community. Underlying much of this work on geographies of organizing is the implicit recognition that workers are constrained in time and in space and that they follow particular time-space paths throughout any particular 24-hour time period – sometimes individuals' time-space paths cross to form *interactive bundles*, i.e. they are in the same place at the same time, whereas on other occasions they remain separate, i.e. people may occupy the same spaces but at different times, or are in different places at the same time, which impacts their capacities for organizing.[3] Recognizing how individual workers' differential ability to cross space at particular times, such that their time-space paths may or may not intersect with those of other workers, can be central to understanding their capacities to organize. For example, the fact that many women have greater domestic responsibilities than do their husbands and so lead lives that are more closely tied spatially to the home than are men's may make it more difficult for them to attend union meetings held late at night (for more on how

3 To my knowledge, no one writing within Labour Geography has drawn explicitly upon the ideas of Swedish geographer Torsten Hägerstrand, who popularized the idea of "time geography" in the 1970s from which such concepts as "interactive bundles" draw. However, the notion that workers and other social actors pursue particular time-space paths throughout the day and that this shapes their capacities for organizing is an important one to recognize (see Hägerstrand [1970], Thrift [1977], and Pred [1981]).

men's and women's domestic responsibilities affect their work and commuting patterns, see Hanson and Pratt [1995]).

Putting all of this together, then, there are several axioms around which Labour Geography has developed:

- social actors are geographically embedded and this shapes the possibilities for their social action;
- for both capital and labour, negotiating the tensions between the needs for fixity and for mobility is a process which drives much of their economic praxis – capital must constantly look for pastures greener even as it must be fixed in place so as to facilitate accumulation, whereas labour must determine whether migrating to new locations is worth abandoning current places of work and residence;
- different sets of social actors are differentially tied into local, regional, national and transnational relationships, and the ways in which they are so will shape their praxis;
- different sets of social actors will often have quite different spatial visions with regard to how the geography of capitalism should be made and these varying spatial imaginations can result in significant political conflicts;
- scale-making is often central to workers' political praxis;
- how social actors behave geographically shapes how landscapes are made, with the result that landscapes are contested social products;
- landscapes are not merely a reflection of social relations but are also constitutive of them – paraphrasing Marx (1852 [1963]), the landscapes made by all the dead generations weigh like a nightmare on the possibilities for action of the living;
- analysing workers' political and economic practice requires an approach grounded in historico-geographical materialism.

Lacunae

Despite the progress made in seeking to understand how workers' praxis shapes landscapes and how the way in which landscapes are made shapes workers' praxis, there are several areas in which Labour Geography probably needs to engage in some critical self-reflection if it is to continue to provide intellectual dividends. The first of these is its empirical focus. Without question, Labour Geography has overwhelmingly focused upon industrial capitalist societies. This is perhaps understandable, given that most Labour Geographers are creatures of the political left who are located in the industrialized capitalist countries of the Global North and take seriously Marx's admonition that the point of analysis is not just to understand the world but, also, to change it. However, it does raise questions about how to broaden Labour Geography's ambit to include analysis both from non-capitalist/non-industrial societies in the present era but also how workers played roles in shaping the landscapes of the past, as with, say, mediæval peasants shaping European feudal landscapes. Related to this, it is fair to say that most of the work within Labour Geography to date has very much seen the landscape in economic terms.[4] Again, this is largely understandable, given that much of the work to which Labour Geographers were reacting, especially the failure to take workers seriously as geographical actors, came out of Economic Geography. Nevertheless, it does mean that aspects of workers' lives other than their worklife tend to have been neglected. Thus, how do workers, as workers, shape not only the spaces within which they work and the connectivities between such spaces, through, for instance, trans-spatial solidarity efforts, but, also, the non-work spaces which facilitate their own self-reproduction? For instance, there is a long history of worker activities within what Marxists would call the sphere of consumption, both informally but also in terms of the creation of organizations designed to represent workers' interests in this regard. Hence, in 1895 members of the Christian socialist movement established the International Co-operative Alliance with the intent of setting up transnational cooperative trading associations (Gurney 1988). However, there has been very little focus upon these kinds of organizations within Labour Geography circles.

Associated with the focus upon workers' economic lives has been a focus upon unionized workers specifically. Once more, there are good reasons for this. Thus, as with capitalists, in working collectively workers are likely to be more capable of shaping landscapes than they are individually or in smaller numbers, and showing examples of workers shaping the landscape of capitalism was important from a theoretical point of view to the early Labour Geography project. Labour Geography, after all, emerged to challenge capital-centric Marxist and other explanatory frameworks. Thus, despite the fact that union membership has been declining in recent decades, the synergy developed by having workers interact with one another within the context of well-defined institutional frameworks means that their impacts on capitalism's economic landscape are likely to be far beyond what their numbers might otherwise suggest. Moreover, many of the early writers associated with Labour Geography had an explicit political agenda – to write the story of the making of capitalism's geography from the point of view of working-class people, and unions clearly represent important bodies designed to facilitate workers' visions within the landscape. There were also practical considerations – unions often have better records and more identifiable

4 This statement, of course, rests upon an ability to distinguish between the economic and the non-economic aspects of workers' lives. This is more problematic than perhaps it may once have seemed, given the influence of post-structuralist thinking that has challenged approaches which have typically seen "the economy" as a "self-evident object of study" (Thrift 2000: 690). Nevertheless, most work within Labour Geography has focused upon workers' paid activities (which are typically, if perhaps not entirely correctly, taken to be the "economic" aspects of their lives), rather than other types of activities within which they might engage.

organizational structures than do other types of organizations, and this can facilitate researchers' investigations into how workers shape landscapes. Nevertheless, now that the basic conceptual arguments have been well accepted within the broader literature concerning workers' spatial praxis, there is a need to move beyond the world of capital L 'Labour' and to develop a wider conception of working-class people as geographical agents, one that moves beyond simply unionized workers. Part of this appears to be occurring with the emergence of what is now being called 'working-class studies' (Russo and Linkon 2005, Mitchell 2005)

An additional lacuna within Labour Geography to date has been that of a general ignoring of the state. Certainly, some scholars have explored how public employees' actions have been shaped by spatial considerations. Painter (1991), for example, suggested that local circumstances have been important influences upon public employees' responses to government privatization in Britain. However, although there has been significant analysis of how capitalists seek to develop particular spatial fixes within the landscape and how workers seek to avoid/rework/outmanoeuvre them, there has been relatively little attention paid to how the state seeks to do so, how workers respond to this, and how their responses in turn shape how the state evolves as a spatial actor. This is significant, given how important the state is as a regulator of workers' lives, whether through workplace-orientated entities, like the US National Labor Relations Board, which adjudicates between employers and unions in collective bargaining disputes, or non-worklife-orientated activities, like local zoning codes, which can affect workers' social and biological self-reproduction practices by regulating how many people may live in a residence. Of course, the fact that the state is also crucial as a more general shaper of landscapes through its investment practices also has implications for workers. It appears, then, that a greater consideration of the state and what its spatial praxis means for workers is timely.

Finally, although much of the conceptual and political impetus for Labour Geography has emerged out of a desire to highlight workers' geographical agency, the concept of agency itself has not been explored in a particularly nuanced manner. As before, it is not hard to understand why this has been so, for capital's agency has likewise not been conceptualized in a particularly sophisticated manner. Capital has simply been seen to have agency to shape the development of capitalism's geography, and Labour Geography has been about exploring how labour, too, has sufficient agency to bring about its own spatial fixes. However, in this regard it is perhaps useful to consider Aristotle's four types of causality which may be at play in any given event, and thereby illustrate agency, these being the material, the efficient, the formal, and the final cause of an action. Ulanowicz (1990: 43) distinguishes these as follows:[5]

> In the familiar example of the building of a house the material cause exists in the mortar, lumber and other supplies going into the structure. The labourers

5 Thanks to Mitch Chapura for bringing this article to my attention.

> and craftsman constitute the efficient cause, while the blueprints, or bauplan, is cited as the formal cause. Finally, the need for housing on the part of eventual occupants is usually taken as the final cause of building the house.

Incorporating a more nuanced view of causality, and therefore of agency, within Labour Geography allows for a more complete working of Marx's dictum that people make their own history, and, Labour Geographers would argue, their own geographies, but not under the conditions of their own choosing. It suggests that different actors may have different structural capacities with regard to each of these four elements, capacities which may change depending upon specific contexts. Such an understanding may provide insights into how contexts can be developed such that workers' capacities to create spaces which they see as emancipatory are enhanced, which is, after all, what the normative side of Labour Geography seeks.

References

Aronowitz, S. 1990. Writing labor's history. *Social Text* 8–9, 171–95.

Barnes, T. 2000. Inventing Anglo-American economic geography, 1889–1960, in *A Companion to Economic Geography*, edited by E. Sheppard and T.J. Barnes. Oxford: Blackwell, 11–26.

Belina, B. and Michel, B. (eds), 2007. *Raumproduktionen: Beiträge der Radical Geography – Eine Zwischenbilanz*. Münster: Westfälisches Dampfboot.

Berndt, C. and Fuchs, M. 2002. Geographie der Arbeit: Plädoyer für ein disziplinübergreifendes Forschungsprogramm [Geography of work: Plea for an interdisciplinary research program]. *Geographische Zeitschrift* 90(3–4), 157–66.

Cairncross, F. 2001. *The Death of Distance: How the Communications Revolution is Changing Our Lives*. Boston: Harvard Business School Press.

Cox, K.R. and Mair, A. 1988. Locality and community in the politics of local economic development. *Annals of the Association of American Geographers* 78(2), 307–25.

Curry, M.R. 1996. On space and spatial practice in contemporary geography, in *Concepts in Human Geography*, edited by C. Earle, K. Mathewson, and M.S. Kenzer. Rowman and Littlefield: Lanham, MD, 3–32.

Friedman, T.L. 2005. *The World is Flat: A Brief History of the Twenty-First Century*. New York: Farrar, Straus and Giroux.

Goodman, R. 1979. *The Last Entrepreneurs: America's Regional Wars for Jobs and Dollars*. New York: Simon and Schuster.

Gordon, D.M. 1978. Capitalist development and the history of American cities, in *Marxism and the Metropolis: New Perspectives in Urban Political Economy*, edited by W.K. Tabb and L. Sawers. New York: Oxford University Press, 25–63.

Gurney, P. 1988. 'A higher state of civilisation and happiness': Internationalism in the British co-operative movement between c. 1869–1918, in *Internationalism in the Labour Movement 1830–1940, Volume 2*, edited by F. van Holthoon and M. van der Linden. London: E.J. Brill, 543–64.

Hägerstrand, T. 1970. What about people in regional science? *Papers of the Regional Science Association* 24, 7–21.

Hanson, S. and Pratt, G. 1995. *Gender, Work, and Space*. New York: Routledge.

Harvey, D. 1978. The urban process under capitalism. *International Journal of Urban and Regional Research* 2(1), 101–31.

Harvey, D. 1982. *The Limits to Capital*. Oxford: Blackwell.

Harvey, D. 1989. *The Condition of Postmodernity: An Enquiry into the Origins of Cultural Change*. Oxford: Blackwell.

Herod, A. 1997a. Labor as an agent of globalization and as a global agent, in *Spaces of Globalization: Reasserting the Power of the Local*, edited by K. Cox. New York: Guilford, 167–200.

Herod, A. 1997b. From a geography of labor to a labor geography: Labor's spatial fix and the geography of capitalism. *Antipode* 29(1), 1–31.

Herod, A. 1997c. Labor's spatial praxis and the geography of contract bargaining in the US east coast longshore industry, 1953–89. *Political Geography* 16(2), 145–69.

Herod, A. 1998. Discourse on the docks: Containerization and inter-union work disputes in US ports, 1955–1985. *Transactions of the Institute of British Geographers, New Series* 23(2), 177–91.

Hudson, R. and Sadler, D. 1986. Contesting works closures in Western Europe's industrial regions: Defending place or betraying class?, in *Production, Work, Territory*, edited by A. Scott and M. Storper. Winchester, MA: Allen and Unwin, 172–93.

Hyman, R. 2004. An emerging agenda for trade unions?, in *Labour and Globalisation: Results and Prospects*, edited by R. Munck. Liverpool: Liverpool University Press, 19–33.

Johns, R. and Vural, L. 2000. Class, geography, and the consumerist turn: UNITE and the Stop Sweatshops Campaign. *Environment and Planning A* 32(7), 1193–1213.

Marx, K. 1852 [1963]. *The Eighteenth Brumaire of Louis Bonaparte*. New York: International Publishers.

Marx, K. 1858 [1973]. *Grundrisse: Foundations of the Critique of Political Economy*. London: Penguin (1993 printing).

Massey, D. 1973. Towards a critique of industrial location theory. *Antipode* 5(3), 33–9.

Massey, D. 1984. *Spatial Divisions of Labour: Social Structures and the Geography of Production*. London: Macmillan.

Mitchell, D. 2005. Working-class geographies: Capital, space, and place, in *The New Working-Class Studies*, edited by J. Russo and S.L. Linkon. Ithaca, NY: Cornell University Press, 78–97.

Mohammad, R. 2010. Gender, space, and labour market participation: The experiences of British Pakistani women, in *Handbook of Employment and Society: Working Space*, edited by S. McGrath-Champ, A. Herod, and A. Rainnie. Cheltenham: Edward Elgar, 144–58.

Ohmae, K. 1990. *The Borderless World: Power and Strategy in the Interlinked Economy*. New York: HarperBusiness.

Painter, J. 1991. The geography of trade union responses to local government privatization. *Transactions of the Institute of British Geographers, New Series* 16(2), 214–26.

Peet, R. 1983. Relations of production and the relocation of United States manufacturing industry since 1960. *Economic Geography* 59(2), 112–43.

Pred, A. (ed.), 1981. *Space and Time in Geography: Essays Dedicated to Torsten Hägestrand*. Lund: C.W.K. Gleerup.

Russo, J. and Linkon, S.L. (eds), 2005. *The New Working-Class Studies*. Ithaca, NY: Cornell University Press.

Savage, L. 1998. Geographies of organizing: Justice for Janitors in Los Angeles, in *Organizing the Landscape: Geographical Perspectives on Labor Unionism*, edited by A. Herod. Minneapolis, MN: University of Minnesota Press, 225–52.

Smith, N. 1984. *Uneven Development: Nature, Capital and the Production of Space*. Blackwell: Oxford.

Southall, H. 1988. Towards a geography of unionization: The spatial organization and distribution of early British trade unions. *Transactions of the Institute of British Geographers, New Series* 13(4), 466–83.

Stenning, A. 2010. Work, place and community in socialism and post-socialism, in *Handbook of Employment and Society: Working Space*, edited by S. McGrath-Champ, A. Herod, and A. Rainnie. Cheltenham: Edward Elgar, 197–212.

Storper, M. and Walker, R. 1983. The theory of labour and the theory of location. *International Journal of Urban and Regional Research* 7(1), 1–42.

Thrift, N. 1977. *An Introduction to Time-Geography*. Catmog 13, Institute of British Geographers.

Thrift, N. 2000. Pandora's box? Cultural geographies of economies, in *The Oxford Handbook of Economic Geography*, edited by G. Clark, M. Feldmann, and M. Gertler. Oxford: Oxford University Press, 689–702.

Ulanowicz, R.E. 1990. Aristotelean causalities in ecosystem development. *Oikos* 57(1), 42–8.

Chapter 3
Re-embedding the Agency of Labour[1]

Neil M. Coe and David C. Jordhus-Lier

Introduction: What is Labour Geography?

Labour Geography can be hard to pin down. It is perhaps best described as a loose coalition of like-minded individuals united by a desire to reveal the multiple geographies that underpin the everyday worlds of work and employment. Chiefly anchored in economic geography, Labour Geographers are also to be found in the sub-disciplines of socio-cultural, political, population and feminist geography. Castree (2008) has identified Labour Geography's five 'signature characteristics' as: a concern with all things geographical; a defence of worker agency; a lack of clear analytical boundaries; an open-minded theoretical sensibility and a Left-leaning political stance. It is possible to discern three distinct strands within the literature, each of which is underpinned by its own bodies of theory. The first has been concerned with the collective organization of workers through labour unions and the reassertion of the potential agency of these groupings (see for instance Herod 1997, Wills 1998). Another has theorized the formation of geographically-specific local labour markets/regimes and their ongoing regulation and segmentation (see for instance Hanson and Pratt 1995, Peck 1996). The third has explored the intersections of employment relations with other facets of personal and workplace identity, most notably gender and ethnicity (see for instance McDowell 1997, Wright 1997). In this context, Peck (2008) has aptly described Labour Geography's internal structure as 'cellular', there being few conversations across the three approaches and with each being more strongly tied to debates and literatures outside the discipline of geography, e.g. industrial relations and working class studies with respect to the first strand.

In this chapter we seek to critically evaluate the contributions of Labour Geography to our understanding of the agency of workers against a backdrop of neoliberal globalization. As such, our analysis, and our use of the term Labour Geography, will largely be confined to the first of the three dimensions described above. Overall, our argument calls for a reappraisal of the claims concerning labour agency that have seemed inherent in what we will shortly characterize as the third phase of Labour Geography. More specifically, we argue that the notion of agency needs to be further conceptualized and fleshed out, and that the potential

1 The same issues raised in this chapter are further elaborated on in *Progress in Human Geography*, April 23, 2010.

for worker action should always be seen in relation to the structures of capital, the state and the community in which labour is incontrovertibly yet variably embedded. The analysis proceeds in three stages. First, we offer a brief account of the development of Labour Geography and highlight the discernible seeds of a more nuanced assessment of worker agency in recent work in this field. Second, we unpack the notion of labour agency along the dimensions of space, time and collective action in a bid to crystallize some of these emergent ideas. Third, we seek to demonstrate the argument that agency is always relational by considering, in turn, labour's positionality with respect to global production networks, the state and the community.

Labour in Geography: Four Phases of Development?

As noted above, Labour Geography is a relatively small, fast-moving and multi-stranded endeavour. Nonetheless, it does, we would suggest, exhibit enough coherence for it to be possible to offer a tentative chronological characterization of its development. We will consider this evolution in four stages: three have been profiled elsewhere (see for instance Herod 1997) and the fourth is arguably in an embryonic stage.

The first phase, which was dominant in economic geography until the 1970s and still resonates both within the discipline, e.g. the regional science approach, and beyond it, e.g. mainstream economics, was based within the traditions of *neoclassical location theory*. The chief concern in this work was to understand how firms made locational decisions, and as such, workers were not theorized in either individual or collective terms apart from as a productive input or locational factor, and on some occasions, as consumers of commodities and services. Workers were bereft of agency, and were presented in terms of categories such as wages, skills levels and gender that could be weighed by firms in their locational decision-making. The advent of *Marxist-inspired economic geography* in the 1970s and 1980s heralded an ontological shift. The path-breaking analyses of scholars such as Harvey, Massey and Smith all highlighted the importance of capital-labour relations and class struggle in shaping the economic geographies of capitalism. That being said, workers were still by and large depicted as an oppressed class that did not have the capacity to directly shape the geography of the capitalist system through their actions. As capital remained the chief shaper of economic landspaces through its locational and investment decisions, Herod has labelled both the neoclassical and Marxist approaches as being concerned with the *geographies of labour*.

The third phase of this intellectual evolution, and one which saw the founding of the Labour Geography sub-discipline as we know it today, came from Herod's clarion call during the 1990s for a shift towards *labour geographies* that would foreground the ability of workers to create and shape economic geographies through pursuing their own spatial fixes. Exposing the weaknesses of the spatial configurations of capital in this way was a political strategy as much as an intellectual one, offering a more optimistic 'corrective to accounts that present workers either as inherently powerless and condemned only to follow the dictates of (global) capital or as simply dupes of capital' (Herod 2001a: 36). A series of case studies emerged that demonstrated the potential of workers to create their own organizations and hamper, or in some cases even overturn, the demands of capital through their actions. Several of these case studies have become iconic within the literature, such as those offered by Herod himself on the upscaling of collective agreements in the US East Coast longshore industry from the 1950s to 1970s, the fight for reinstatement by locked-out metalworkers at the Ravenswood Aluminium Corporation plant in West Virginia in the early 1990s, and the disruption of General Motor's Just-in-Time production systems by workers in Flint, Michigan in 1998 (compiled, with other studies, in Herod 2001a).

Through such work, recognition of the range of spatial strategies available to workers was enhanced. Campaigns could be locally or translocally organized, and the targets of such solidarity in turn could be either locally or non-locally based. Importantly, the work revealed that labour internationalism, whether formally coordinated or grass-roots in nature, was but one of many different strategies available to workers. That being said, this early wave of labour geographies was not without its limitations (Lier 2007). First, there was a focus on labour union activity at the expense of worker agency not articulated through collectively organized strategies. Second, there was a seeming bias towards isolated success stories of workers with particular capacities to act and enhance their relative position *vis-à-vis* capital. Third, in sectoral and geographical terms, most focus appeared to be on manufacturing sectors in developed countries, perhaps an inevitable reflection of the intellectual origins of Labour Geography itself. Fourth, and particularly pertinent in the context of this chapter, the notion of labour agency remained curiously under-theorized (Castree 2008). We shall return to this issue shortly.

In recent years, however, it has perhaps become possible to discern what might be thought of as a fourth phase in the evolution of Labour Geography which has sought to tackle these issues. We do not want to overstate the case here – the fourth phase is, like its predecessor, essentially a loose grouping of case studies and does not offer a unified analytical framework. Moreover, drawing a precise temporal dividing line between the phases is difficult. And, indeed, some may simply prefer to think of this latter work as an internal evolution of the labour geographies phase. However, for us there are some exciting tendencies within this newer body of work that make it worth isolating as a distinct stage. In short, the remit of labour geographies seems to have extended in novel and overlapping directions to incorporate *new sectors* such as low-paid service and public sector work (see for instance Aguiar and Herod 2006), *new domains of action* such as consumption-based campaigns (see for instance Hartwick 1998, Johns and Vural 2000), *new modes of organization*, and in particular, community-oriented unionism (see for instance Wills 2001, Lier and Stokke 2006) and *new geographies* such as post-Apartheid South Africa and the subcontracted sweatshop zones of developing countries among many others (see for instance Lambert and Webster 2001, Hale and Wills 2005). Although agency is rarely discussed explicitly, as they have ventured further beyond the core workers of core countries these studies also hint at a more ambivalent and less sanguine attitude to the potential for, and of, worker action of different kinds. It is this aspect that we wish to explore in the remainder of the chapter against a real-world backdrop that has arguably, for many at least, seen a continued erosion of worker rights and opportunities over the years that Labour Geography has emerged and evolved. What does it really mean to talk of worker agency in the harshly competitive context of what Jane Wills (2008b) calls 'subcontracted capitalism'?

Rethinking Agency

Labour agency, then, is a key concept in Labour Geography, and one which is central to defining the purpose of the sub-discipline *vis-à-vis* other geographical perspectives on the world of work. Still, coming to terms with agency is not easy. In simple terms, Labour Geographers seem to be asking: what can labour do, besides work? Can workers actively shape the world they live in? In other words, *does labour agency matter* in a global capitalist economy? An unanimous *yes* to the latter question seems to give a certain coherence to the field of Labour Geography, and numerous case studies have been brought forward to illustrate the many ways in which worker action matters. While these contributions have been important, both theoretically and in a more political sense, it becomes clearer, as the field of Labour Geography continues to grow, that there are limitations to the usefulness of promoting the potential of labour agency *in general*. While Labour Geographers have been very effective in exploring this potential through case studies, the diversity of these case studies also implies that labour's room for manoeuvre is uneven.

This unevenness, in turn, plays out across some key dimensions. We suggest that clarifying these through a tentative typology would be a useful exercise, as it can take our discussion from for-or-against labour agency to a more detailed understanding of labour agency as variegated.

A first dimension could be labelled the *geography of agency*. In a way, that is what Labour Geographers have been grappling with, more or less consciously, and it is not the task of this chapter to give an exhaustive account of the many ways which labour agency is geographical. But without stating the obvious, some points are worth mentioning. One of the most helpful contributions to this understanding is the use of *political scale*. Scale is particularly useful in that it captures the double nature of the spatiality of labour agency. Not only do the actions of labour play out in complex social geographies, but they can be understood as spatial phenomena in themselves. In other words, both the conditions and the strategies of labour agency are spatial. Moreover, scale has been used by labour to reconfigure political landscapes and renegotiate social hierarchies in ways which are more beneficial to the interests of workers. The focus on political scale directs attention to actors beyond the labour-capital nexus, and shows how regulatory institutions are crucial to the politics of labour. Wills' (2008a) approach to the potential of agency, *mapping class*, is another welcome intervention. While Labour Geography emerged in part as a response to a capital-centric bias in the literature, it has not yet contributed much to an analysis of class. But as Wills illustrates, when we try to map the working class, e.g. according to workers' relation to the production of surplus value, we discern a landscape of highly uneven political possibilities. In summary, geography is fundamental to understanding agency and to defining labour's potential as a political actor.

Interestingly, there is also a lot to be learnt about the temporal aspects of union organizing from reading case studies in Labour Geography, although this point is often understated in the literature. Historical time, of course, is important, but this has been well covered by the field of labour history and economic approaches such as regulation theory. A different angle from which to approach the temporal dimension is to look at time-as-process which, in many cases, is of crucial explanatory significance. Firstly, it is important to understand the *timing of agency* and the dynamics between particular union struggles in relation to external circumstances. Herod (2001b), for example, acknowledges that the success of the Ravenswood campaign to a certain extent relied upon the national umbrella body AFL-CIO championing it as a *cause célèbre*. In other words, the prospects to succeed for labour struggles located in similar spatial-economic hierarchies in the US could possibly depend on whether they occurred before or after Ravenswood. Secondly, the temporal dimension is also important with regards to the internal dynamic of labour mobilization. Moments of agency form part of a *learning process*, and should be understood as such. The point made here is that unionism is based on trial-and-error, and a reorientation of strategies might need time to develop in place before being politically effective. Thirdly, *labour agency often evolves in a dialectic relationship to external processes* such as economic restructuring or political reform. One of the authors' case study of the politics of the South African Municipal Workers' Union (SAMWU) in relation to local state restructuring in Cape Town exemplifies this point. After an anti-privatization campaign in the late 1990s, which drew on the organization's tradition of social movement unionism during apartheid, the union had to start dealing with the actual results of outsourcing in the 2000s, e.g. by launching recruitment drives in private companies. As political opposition proved insufficient to stop neoliberal change, the union also attempted to play the role of a social partner in corporatist forums. These phases can in part be understood as an internal organizational learning process, but they also represent labour adapting to changes in local political regimes and in the urban labour market. By focusing on isolated moments of agency, these dynamics are hard to spot. In the same vein as Labour Geography can be criticized for a bias in terms of sector and geography, a related concern can be voiced with regards to selectivity in time. How do workers act in the period following victorious struggle, for example?

Another oft-mentioned bias in the literature is the tendency to conflate worker agency with union agency. Trade unions are not the only active labour agents in the making of economic geographies. Cumbers et al. (2008) are right in arguing that it is important to distinguish between abstract labour as an analytical category and specific forms of union organizing as political phenomena. Not only should other organizations of workers be more fully recognized but, in relation to that, other subject positions on which workers act must be better understood in our writing of labour geographies: workers as citizens, consumers and or as family members. Even within the workplace, their loyalties and identities are riven between their occupational roles, their interests as wage earners and their union affiliation. All these subject positions may lead workers to act as spatial agents, sometimes in contradictory ways.

But how do we distinguish between trade union politics and other forms of labour agency? Can binaries such as *individual versus collective, or political versus non-political* bring us further? Labour migration is a phenomenon which challenges these binaries in interesting ways. It exemplifies how spatial agency is not dependent on trade union organization, and that both individual and collective labour agency can be explicitly spatial. But while migrating in search of work is often thought of as an individual or inter-personal act, it is often collectively organized and leads to broader demographic patterns. The process of labour migration can hardly be described as non-political either, as it changes the political economy of the area of origin and lead to new political cultures in the places where migrants settle down (de Haan and Rogaly 2002). But even if we take the conscious decision of limiting our discussing to collective agency in the form of trade union action, some distinctions are worth making. Webster et al. (2008) provide a typology of collective agency, albeit indirectly, through their discussion of power. And the relational nature of agency does indeed remind us not to exclude power from our analysis. Drawing on Marxist class analysis, they suggest that we can distinguish between the structural power that workers display when they halt an economic system through downing their tools, and the associational power represented by the organizational strength of a political labour movement. Interestingly, while the first form of power is based on suspending a certain action, not working, the other is based on workers actions through political mobilization. Again, the potential of both these kinds of collective agency is extremely uneven between different groups of workers, and the ways in which they are embedded in economic systems, political regimes and social networks.

Re-embedding Agency

It becomes clear, from the discussion above, that labour agency can not be treated as a coherent phenomenon in the world economy without gross overgeneralizations. We argue that once Labour Geography has established the very important rhetorical and theoretical point that labour can make a difference, it is time to move one step further and analytically examine and conceptualize the relationships between labour and other social actors, institutions and networks. While the embeddedness of labour agency can be framed and understood in different ways, we suggest that three social structures are fundamental to labour, both as a social category and as a set of political organizations. These can be conceptualized in their most general terms as capital, the state and the community. In what follows we will discuss each of these briefly, and outline ways to approach labour's embeddedness in each of them.

A key starting point for a reassessment of labour agency has to be the changing nature and scale of the organization of capital. Our preferred term for characterizing the increasingly globally-integrated nature of contemporary production systems is *global production networks* (GPNs). GPNs can be thought of as the globally organized nexus of interconnected functions and operations of firms and non-firm institutions via which goods and services are produced and distributed. Through the processes of economic globalization, such networks have become far more organizationally complex and geographically extensive. Moreover, GPNs are inherently dynamic and are always in flux in response to both internal and external circumstances. In common with the first and second phases of Labour Geography literature described above, the GPN literature, and cognate work on global commodity/value chains, has remained notably silent on the issue of labour agency. Labour is, most commonly, simply assumed to be an intrinsic part of the production process and workers are typically presented as passive victims of capital's inexorable global search for cheaper wages. This is a particularly stark omission given that GPNs are as much networks of embodied labour as they are systems of value creation (Cumbers et al. 2008).

Enhanced dialogue between the GPN and Labour Geography literatures would seem to offer benefits to both sides. From the GPN perspective, analytical space needs to be created for both the possibilities and realities of labour agency (Coe et al. 2008). Its currently rather firm/capital-centric approach needs to at least be tempered by the recognition that some workers have the agency to improve their relative position and, at the same time, to contribute towards re-shaping economic geographies. More important in the current context, however, is our assertion that Labour Geography can benefit from the analytical insights offered by the GPN perspective. In seeking to reveal the fragmented yet tightly coordinated organization of capital at the global scale, GPN analysis can simultaneously serve to reveal the *variegated landscape for agency potential* across different sectors. While some workers occupy privileged positions within the broader system, associated, for example, with high levels of value adding activity and low levels of potential substitutability, others find themselves in far more marginal, transitory and competitive working environments. The key point here is that there are massively different levels of potential agency *within* functionally-integrated economic networks. Adopting this kind of analysis is productive in at least three respects. First, it serves to reveal weak spots within the overall system that workers may be able to exploit, for example, logistical choke points, areas of potential exposure to consumers or a high dependence on workers with unique skills. Second, potential lines of solidarity between different groups of workers in different places are revealed. On the other hand, thirdly, the potentially deleterious impacts of worker action on other groups of workers within the same functional system, otherwise known as geographical dilemmas (Castree et al. 2004), can be brought into view. In short, adopting a GPN-informed perspective allows the potential for labour agency to be interpreted and evaluated in the context of the wider global economic system of which workers are a part.

Reconnecting worker agency to the structures of capital in this way also affords a potentially more realistic, if somewhat depressing, perspective on the limits to worker action in the contemporary global system. As Rutherford and Holmes (2007: 196) argue, 'the emphasis placed on labour's agency by some labour geographers needs to be tempered by considering the continued significance of macro-processes'. As noted earlier, the Labour Geography literature tends to focus on those employees within GPNs whose position offers them the potential to exert effective pressure on their employers. The reality for many workers, however, is very different. It is difficult to consider upscaling worker action, for example, without pre-existing local/national organizational structures (Lier 2007). The fundamental spatial asymmetries between labour and capital, based upon the relative fixity of labour and the greater mobility of capital, remain a powerful limiting condition on effective mobilization. The fact that, globally, the level of labour force unionization has continued to decline, albeit unevenly, and that the share of income going to labour has also continued to decline whilst, at the same time, the effective global labour supply quadrupled between 1980 and 2005, is a salient reminder of the wider conditions here.

Positioning labour within GPNs represents one way in which labour is embedded within the sphere of production. Importantly, however, labour's ability to act is also conditioned by the spheres of regulation and reproduction. The focal point of the sphere of regulation, the *state*, remains the institutional apparatus which regulates the lives and politics of workers. An analysis of labour and the state elaborates arguments made earlier in this chapter: the state takes on multiple roles *vis-à-vis* workers and their representative organizations; labour, in turn, engages the state in all of these capacities, sometimes at once. From an analytical point of view, however, some of these roles have received more attention than others. In geography, the *state-as-regulator* has in particular been thoroughly examined through the regulation of labour markets. Also, state territorial units have been conceptualized as *containers of labour practices*. But other roles of the state, e.g. being a *provider of basic services*, are also of critical importance to working-class politics. Finally, the state is also a *political apparatus* and an arena where practices of domination and democratic influence take place. As Labour Geographers, all of these roles must be accounted for if we are to understand how labour agency is embedded in state power and practices.

If we turn our attention towards public sector workers, a still-prominent part of the labour force in most industrialized countries, we also have to attribute to the state the task of *employer and boss*. The politics of public sector unions are illustrative of the multiple roles which both labour and the state take on in relation to each other. The public sector arguably presents workers with some unique challenges and opportunities. In contrast to workers in competitive sectors with a high degree of capital mobility, which calls for transnational labour mobilization, state employees are still dominated by dynamics at the national scale. This confines the agency of organized labour based in the public sectors (of national economies) and their prospects of up-scaling. At the same time, neoliberal state restructuring forces organized labour in this sector to grapple with increased competition between local state authorities and the erosion of the sector as we know it, through processes of externalization and privatization. The fact that their employer, the state, is a political and democratic institution, however, can possibly increase the agency potential of public sector workers. But this necessitates that groups of workers accrue what Webster et al. (2008) call 'symbolic power', i.e. organizing around issues which attract sympathy from the general public to which democratic authorities are accountable. Most public sector workers reside in areas dealing with public services on a day-to-day basis, and they are concerned as users of these services. Because services produced in the public sector are often basic services produced for collective consumption, the potential for forging solidarity links between organized labour and other groups of citizens is particularly promising for certain groups of public sector employees. The flipside of this, though, is the tendency by state authorities to define the health and service delivery sectors as 'essential services', thereby legally excluding these workers from organizational rights such as the right to strike. The relationship between public sector workers and the state is complex, and organised labour is able to exert pressure on its employer through different approaches. Still, we should be careful not to interpret this multitude of roles and relationships as increased agency. Although it clearly offers alternatives, juggling different hats can also be counter-productive, or even paralysing, for trade unions in the public sector.

The relationship between organized labour and other social constituencies lead us to the sphere of reproduction. The *community*, and the localized social networks in which all workers are embedded, represents the third structure in which we aim to re-embed our understanding of labour agency. This is an important point, and one which Labour Geographers have picked up on in different ways. Not only are we starting to recognize other worker subjectivities than those based in class or profession, but there has also been increased interest in the many non-workplace issues which mobilize workers. In this respect, Labour Geography has come some way in responding to Castree's (2008) call for a holistic understanding of the worker. This, in turn, opens for other collective labour agents, such as non-union organizations and coalitions between trade unions and other groups. This is particularly relevant when we deal with processes of economic restructuring and deregulation that lead to increasing insecurity for workers. When social risk, which hitherto had been mediated by employment contracts and welfare state intervention, is left with the workers, they must rely on the social networks they are part of. Put in simple terms, labour comes to rely on the community when the community of the workplace is put at risk.

One key challenge facing many workers in the age of neoliberalism is posed by a set of spatial dynamics which can be loosely grouped under the concept of *fragmentation*. Fragmentation relates to how, where, when and by whom work is organized. New forms of workplace and work-time organization such as outsourcing and subcontracting of services are often the results of employers seeking to increase productivity, reduce costs and combat the power of organized labour. The consequence for workers is often a multi-dimensional fragmentation of the workforce: not only across space, but between a network of direct and indirect employers, through customization and individualization of contracts, and 'in time' through shift-work and short-term engagements. What all these processes have in common is that they represent an obstacle to building working-class solidarity and make unionism in the traditional union-shop sense less effective. But these new geographies of work have triggered responses from innovative labour agents. Many of these responses have been documented in the case study-based literature on community-oriented unionism, and they provide a telling illustration of how labour agency is embedded in the community (Savage 1998, Tufts 1998, Walsh 2000, Pastor 2001, Wills 2001). We can conceptualize community-oriented unionism as a redrawing of geography in its own right, as a socio-spatial response to fragmentation. Firstly, some of these movements *organize around new scales* where the alliance between groups of workers and other social constituencies are strong, from local neighbourhoods to the city region. Secondly, community-oriented unions have been able to *identify alternative targets for action* in a socio-economic landscape where agency is particularly constrained in relation to the employer. Confronting retailers, consumer groups or local state authorities might represent this strategy. Thirdly, unions have used residential areas and community centres as *sites of recruitment*, when the workplace is fragmented or inaccessible. In sum, these represent some of the ways in which organized labour might use their embeddedness in the community to their political advantage.

Conclusion

The dynamics of capital accumulation have, particularly through processes of neoliberalization, worked to fragment labour agency as a political project. Our ambition in this chapter is not to further this fragmentation of labour agency at a conceptual level. Rather, we want to identify ways of maintaining explanatory power amidst this complexity. There have been two inter-linked threads to our argument. The first is that we need to be more precise and systematic in our use of the term *agency* and how it is facilitated or proscribed by geographical processes. By starting to unpack agency along the dimensions of geography, time and collective action we have revealed how this might be analytically beneficial, but much remains to be done in this regard.

Our second argument has been that to simply assert the apparently free-floating agency of labour is never enough. Labour agency is a relational concept that can only ever be understood in the context of the structures of capital, state and community in which all workers are inevitably but differentially embedded. The multiple subject positions of workers means they constantly operate within complex and variable landscapes of opportunity and constraint. For some, their positionality within these structures will offer considerable 'wiggle room'. For others, however, meaningful agency will be tightly circumscribed by the intersection of profound structural forces. Re-embedding the agency of labour in this way is, we would posit, extremely important to advancing the project of Labour Geography theoretically, methodologically and politically.

References

Aguiar, L.L.M. and Herod, A. (eds), 2006. *The Dirty Work of Neoliberalism: Cleaners in the Global Economy*. Oxford: Blackwell.

Castree, N. 2008. Labour geography: A work in progress. *International Journal of Urban and Regional Research* 31(4), 853–62.

Castree, N., Coe, N., Ward, K. and Samers, M. 2004. *Spaces of Work: Global Capitalism and Geographies of Labour*. London: Sage Publications.

Coe, N., Dicken, P. and Hess, M. 2008. Global production networks: realizing the potential. *Journal of Economic Geography* 8(3), 271–95.

Cumbers, A., Nativel, C. and Routledge, P. 2008. Labour agency and union positionalities in global production networks. *Journal of Economic Geography* 8, 369–87.

de Haan, A. and Rogaly, B. 2002. Introduction: Migrant cultures and their role in rural change. *Journal of Development Studies* 38(5), 1–14.

Hale, A. and Wills, J. 2005. *Threads of Labour*. Oxford: Blackwell Publishing.

Hanson, S. and Pratt, G. 1995. *Gender, Work and Space*. London: Routledge.

Hartwick, E. 1998. Geographies of consumption: A commodity-chain approach. *Environment and Planning D: Society and Space* 16, 423–37.

Herod, A. 1997. From a geography of labor to a labor geography: Labor's spatial fix and the geography of capitalism. *Antipode* 29(1), 1–31.

Herod, A. 2001a. *Labor Geographies: Workers and the Landscapes of Capitalism*. New York: Guilford Press.

Herod, A. 2001b. Labor internationalism and the contradictions of globalisation: Or, why the local is sometimes still important in a global economy. *Antipode* 33(3), 407–26.

Johns, R. and Vural, L. 2000. Class, geography and the consumerist turn: UNITE and the Stop Sweatshops Campaign. *Environment and Planning A* 32, 1193–1214.

Lambert, R. and Webster, E. 2001. Southern unionism and the New Labour internationalism. *Antipode* 33(3), 337–62.

Lier, D.C. 2007. Places of work, scales of organising: A review of labour geography. *Geography Compass* 1(4), 814–33.

Lier, D.C. and Stokke, K. 2006. Maximum working class unity? Challenges to local social movement unionism in Cape Town. *Antipode* 38(4), 802–24.

McDowell, L. 1997. *Capital Culture: Gender at Work in the City*. Oxford: Blackwell.

Pastor, M. 2001. Common ground at Ground Zero? The new economy and the new organizing in Los Angeles. *Antipode* 33(2), 260–89.

Peck, J. 1996. *Work-place: The Social Regulation of Labor Markets*. New York: Guilford Press.

Peck, J. 2008. *Labor Markets from the Bottom-up*. Presentation to the 4th Summer Institute in Economic Geography. University of Manchester, 13–18 July 2008.

Rutherford, T.D. and Holmes, J. 2007. 'We simply have to do that stuff for our survival': Labour, firm innovation and cluster governance in the Canadian automotive parts industry. *Antipode* 39(1), 194–221.

Savage, L. 1998. Geographies of organizing: Justice for Janitors in Los Angeles, in *Organizing the Landscape: Geographical Perspectives on Labor Unionism*, edited by A. Herod. Minneapolis, MN: University of Minnesota Press, 225–53.

Tufts, S. 1998. Community unionism in Canada and labour's (re)organizing of space. *Antipode* 30(3), 227–50.

Walsh, J. 2000. Organizing the scale of labour regulation in the United States: Service sector activism in the city. *Environment and Planning A* 32, 1593–1610.

Webster, E., Lambert, R. and Bezuidenhout, A. 2008. *Grounding Globalization: Labour in the Age of Insecurity*. Malden, USA: Blackwell Publishing.

Wills, J. 1998. Taking on the CosmoCorps? Experiments in transnational labor organization. *Economic Geography* 74(2), 111–30.

Wills, J. 2001. Community unionism and trade union renewal in the UK: Moving beyond the fragments at last? *Transactions of the Institute of British Geographers* 26, 465–83.

Wills, J. 2008a. Mapping class and its political possibilities. *Antipode* 40(1), 25–30.

Wills, J. 2008b. Subcontracted employment and its challenge to labor. *Labor Studies Journal* OnlineFirst, published 15/10/08 as doi:10.1177/0160449X08324740.

Wright, M.W. 1997. Crossing the factory frontier: Gender, place and power in the Mexican maquiladora. *Antipode* 29(3), 278–302.

PART II
The Agency of Unions

Chapter 4
The Entangled Geographies of Trans-national Labour Solidarity

Andrew Cumbers and Paul Routledge

Introduction

With the end of the Cold War in 1989, there was considerable hope that the international trade union movement would move beyond its traditional political and ideological schisms to begin to construct more genuine relations of transnational solidarity between workers. Certainly, there was a pressing need for new forms of global labour action to counteract the globalization of capital and the emergence of ever more complex networks of production and exchange (Dicken 2007), and during the 1990s a lively debate ensued about the forms that a more progressive and grassroots trade union internationalism might take (Moody 1997, Waterman 1998, Waterman and Wills 2001). Two decades on, a charitable view would be that progress towards genuine trans-national solidarity has been rather pedestrian. A more critical one would be that some old divisions, particularly between north and south, remain as entrenched as ever, whilst some of the new forms of global union organization are largely irrelevant to the challenges facing workers (see for instance Waterman 2007).

Despite such reservations, trade unions are attempting to construct new forms of trans-national solidarity. The material realities of an increasingly global economy are forcing unions to recognize the importance of trans-national organizing and campaigning, resulting in a number of new initiatives in the period since the mid-1990s. In this chapter, we explore the complicated politics and geography of trans-national labour solidarity through the lens of a global union federation (GUF), the International Chemical, Energy, Mining and General Workers Federation (ICEM).

Our approach here emphasizes the importance of a geographical understanding to unpacking the strategies and practices of ICEM and its affiliate unions. In particular a set of *geographical entanglements* continue to infuse union operations reflecting local and national interests against broader visions of solidarity. While promoting transnational labour rights is an aspiration of most union actors, this is inevitably compromised by different subject positions both within the spaces of the international labour movement and in relation to broader processes of capital accumulation.

Following the Introduction, the second part of the chapter develops our conceptual argument through the use of the term *entangled geographies* to signal the way space and place mesh to shape the political possibilities and constraints of transnational labour solidarity. Spatially, the global connections of union organization open up possibilities for labour solidarity, but these are themselves still heavily contingent upon place-based, though not bounded (see Featherstone 2008), social actors. The third section of the chapter develops this theme, in respect to the strategies of ICEM, to demonstrate the weakness of top-down models of solidarity shaped by the geographical and political traditions formulated in particular places and imposed elsewhere. The fourth section extends this critique by highlighting the importance of more locally based and grassrooted forms of transnational solidarity and points to some positive examples. The conclusion then argues for decentred forms of union action that can create mutual affinities and more reciprocal forms of solidarity by combining the diverse experience of struggles in particular places with the common experience of alienation and class politics under global capitalism.

Trans-national Solidarity and the Entangled Geographies of Labour Internationalism

In the face of deepening processes of economic globalization, the 1990s witnessed a wave of rhetoric from union leaders espousing the cause of transnational labour solidarity. Even right wing union leaders such as Bill Jordan, the former General Secretary of the International Confederation of Free Trade Unions, championed the cause of internationalism: 'International solidarity must become a natural reflex throughout the union movement' (Bill Jordan 1996, cited in Munck 2002: 13). The loss of union influence through the transfer of work from organized workplaces in the global north to new production zones in the global south was having a deleterious effect upon union power and influence in the global economy. Of the estimated 2.23 billion waged workers in the global labour force, only 295 million (13 per cent) were unionized in the year 2000 (figure derived from Castree et al. 2004: 11 and ICTUR 2005: 5–7).

The extent to which such rhetoric has been transformed into a more progressive agenda for trans-national solidarity is another matter. In reality international union networks continue to be infused with the same entangled geographies that have frustrated internationalism in the past. We use the word *entangled* here to reflect the way international union strategies are bound up in, and complicated by, a series of connections and power relations that cut across national and trans-national scales of union organization.[1] The adjective *entangled* also signals the meshing of space and place that is critical in understanding the possibilities for progressive

1 There is not the space here to develop this argument in depth so we refer the reader to Cumbers et al. (2008) and Routledge and Cumbers (2009) for greater development of this argument in relation to broader Global Justice Networks and the associated concept of convergence space.

forms of transnational solidarity within transnational labour networks.[2] Existing and intensifying transnational connections, both between the leadership and grassroots of the labour movement, offers real possibilities for more genuine forms of transnational solidarity to emerge against the backdrop of an increasingly integrated global economy. Hence, the broader spatial and relational networks that shape union action at local and national scales, need to be recognized over more bounded conceptions of place (Massey 2004). Nevertheless, it is important to recognize the continuing influence of territorially embedded institutions and structures in shaping power relations (Routledge and Cumbers 2009).

In the case of the labour movement, effective power over decision-making and the mobilization of resources for international union action continues to reside at the national scale. Trans-national organizations such as the Global Union Federations (GUFs) and the International Federation of Trade Unions remain largely dependent on national union organizations for their resources. It is the latter who have the membership base and finances to support international activities. In this sense, there are no global unions as such but relatively loose trans-national networks of affiliated national unions. More often than not, the latter prove unwilling either to delegate decision-making power and resources down to the grassroots that might enable workers to forge horizontal networks of solidarity within the operational spaces of MNCs, or to transfer power to the international scale to allow more effective global responses to MNC actions (Cumbers et al. 2008a).

Second, what we have termed elsewhere the *convergence spaces* (Routledge 2003, Cumbers et al. 2008b, Routledge and Cumbers 2009) of transnational union action are highly uneven, reflecting the differential power relations between national union organizations. While most unions have suffered from declining membership and influence in national membership in the face of globalization and neoliberalism, this varies considerably between countries. In the Anglo-Saxon economies, such as the US, the UK, Australia, and New Zealand, unions have been considerably weakened in relation to their own governments and in dealing with business interests but they continue to enjoy legitimacy as social partners in Nordic and continental European states. Even in France, where union formal membership is very low, rights to employee representation in corporate governance structures provides unions with political leverage and therefore the ability to exert pressures on the activities of French MNCs abroad. In the global south, union influence at the national scale is more restricted. In Brazil and South Africa, union strength and legitimacy nationally has grown over the past three decades, tied to the union role in broader social and democratic change outside the workplace. In many other countries, however, unions are non-existent, persecuted or outlawed by their states.

This uneven cartography of union power results in some actors becoming more empowered than others in forging transnational agendas and discourses, leading

2 This differentiates the term from its earlier (but related usage) in thinking about power relations in contemporary societies (Sharp et al. 2000).

to considerable tensions in global union networks (Routledge and Cumbers 2009, Cumbers et al. 2008b). Typically, as we demonstrate below, this can result in the scaling up of particular national models of union-employer relations, which are applied in other geographical contexts irrespective of local conditions and existing sets of social relations. A key faultline continues to be the imposition of top-down models of trans-national labour organizing from unions in parts of the global north to the global south at the expense of the development of more reciprocal forms of trans-national solidarity.

Top-down Solidarity: The Uneven Convergence Space of ICEM[3]

The ICEM (International Federation for Chemical Energy Mine and General Workers) is one of the ten Global Union Federations (GUFs), formerly known as International Trade Secretariats set up to represent and coordinate worldwide labour interests. Nominally, it is a more global organization than many of the MNCs that it deals with, incorporating 408 member trade unions from 125 countries representing 20 million workers, but in reality its power is more circumscribed. Core funding for international work comes from affiliation fees from national affiliates, so that it exists in a dependency relationship with a few powerful national union centres.

ICEM operates through a top-down or *verticalist* logic, displaying a conventional hierarchical structure, centred upon powerful national affiliates, with a leadership, mass membership, and social relations based on delegation, and formal organizational processes. The powerful and largely right-wing German *Industriegewerkschaft Bergbau, Chemie, Energie* (IG-BCE) is the largest contributor to the ICEM.[4] The second most important affiliate, in terms of members and voting power, is the US union PAICE (Paper and Allied Industries, Chemical and Energy Union). Neither of these unions has a history of international work, but, along with key Japanese affiliates, they exercise considerable power in blocking initiatives if they are perceived as contrary to national union interests. Hierarchal and uneven relations within ICEM are reinforced by its operational mechanisms. For example, the statute that 'Only persons holding a post in an

3 The following sections draw upon fieldwork based on 47 interviews with union officials from ICEM and its national affiliates undertaken in 2005 and are derived from a larger project on Global Justice Networks (see Routledge and Cumbers 2009). Six interviews were conducted with ICEM staff and officials involved with the global headquarters (Belgium), including interviews with the current General Secretary and his predecessor; with the other interviews being with trade unionists from national affiliates in Germany (2); the UK (17); France (14); and Norway (8). Two group discussions were also conducted with 'grassroots' members of ICEM affiliate unions in the UK and France.

4 Though it has not been possible to get a complete breakdown of voting numbers, figures given during an interview provide some indication of relative voting strengths with the German affiliate having the largest bloc of 140, as against 80 for the Japanese affiliates and 60 for UK unions (Authors' interviews).

affiliated trade union shall qualify for an elected position' (ICEM 2003: 8) excludes many grassroots members. The voting structure whereby each affiliate union is entitled to one vote for every 5,000 members (ICEM 2003) tends to reinforce the domination of larger, established European affiliates.

Most of the transnational solidarity work is carried out by professional officers and elected national level officials to the extent that grassroots engagement is generally lacking. In practice, a few key individuals, who we term *imagineers* (Routledge and Cumbers 2009), are critical in the development of ICEM's global vision and strategy. Critically, these imagineers emerge predominantly from European and North American unions, bringing particular perspectives and worldviews with them. This relational topography of key social relations spanning the national and trans-national scales is critical to understanding how particular national union political perspectives become hegemonic in global union discourses and practices.

Since 1999, the pillar of ICEM strategy has been the signing of Global Framework Agreements (GFAs) with MNCs (Table 4.1). GFAs are voluntary agreements signed with multinational companies, which pledge to respect a set of principles on labour and trade union rights, including the rights for unions to organize, equality of conditions between genders and ethnic groups and the prohibition of child labour. They are not legally enforceable in either national or international law and as such do not have the status of proper collective bargaining agreements although that is sometimes the perception by workers in the global south. Usually, this reflects a lack of information and communication from the union in the home country, which is itself a manifestation of the unequal power relations that exist in the global union movement (Authors' interviews). The first such agreement was signed in 1999 with Statoil, a Norwegian oil company and subsequently another 12 have been signed. The willingness to sign GFAs reflects the commitment to a partnership approach to industrial relations within ICEM that marks it out as less adversarial in its attitude to MNCs than some other GUFs.

Labour researchers and trade unionists display mixed feelings about GFAs. On the one hand, there are concerns that they legitimize MNC actions with a 'tick box' corporate social responsibility without any legally binding commitment to improve workers' rights (Gibb 2005). However, a well-organized union can use considerable pressure through the media and consumer campaigns to help enforce agreements and highlight abuses. At the more operational end, GFAs do provide a framework for unions to organize at the global level. Certainly, they are no panacea either for guaranteeing labour rights or for developing independent labour organization, and their usefulness ultimately depends upon a GUF's own internal organization and political practices to exploit the opportunities that they offer.

In this light, our evidence from ICEM suggests mixed results so far. The ICEM-Statoil agreement has already been used to support unions in struggles with employers in Nigeria, Poland and the United States as well as helping to build new local union organizations in Azerbaijan (Cumbers 2004, Gibb 2005), although the independence of this union from state and corporate interests in the latter case has

Table 4.1 Global Framework Agreements signed by ICEM and affiliates as of October 2007

Company	Employees	Country	Sector	No. of countries involved	Year
Statoil	16,000	Norway	Oil	23	1998
Freudenberg	27,500	Germany	Chemical	41	2000
Endesa	13,600	Spain	Power	12	2002
Norske Skog	11,000	Norway	Paper	14	2002
AngloGold	64,900	South Africa	Mining	9	2002
ENI	70,000	Italy	Energy	67	2002
RAG	37,000	Germany	Energy and chemicals	not known (n.k.)	2003
SCA	46,000	Sweden	Paper	sales in 90, production in 40	2004
Lukoil	150,000	Russia	Oil	mainly Russia with refining operations in E.Europe, marketing in US	2004
Electricite de France	167,000	France	Energy	17	2005
Rhodia	20,000	France	Chemical	45 sites outside Europe in North and Latin America, Asia-Pacific	2005
Lafarge	77,000	France	Building materials	n.k. but active in Europe, North and Latin America, Asia-Pacific	2005
Umicore	14,000	Belgium	Advanced materials		2007

been questioned by some of our respondents (Author's interviews). An important innovation in the recent deal with French company, Rhodia, was the agreement to work with NGOs, acknowledging that the latter are often better placed to work with local actors and monitor the extent to which MNCs are abiding by agreements. ICEM has also extended the global reach of GFAs by being the first GUF to sign agreements in Russia and South Africa.

However, there is little evidence thus far that the ICEM's GFAs have been successful in building new union organization at the local level. In Gibb's (2005) recent research into the impact of GFAs, it is the metalworkers federation (IMF) and the construction workers (IFBWW) who elicit praise for developing a local union presence in new regions and with subcontractors, rather than ICEM. Critically, the lesson from Gibb's work seems to be that it is only where attention has been paid to the scalar dynamics of union relations, especially in fostering close and open connections between global union federation, national affiliates and local activists that new and vibrant grassroots organizations can develop. In contrast, our research suggests that the ICEM leadership in Brussels has attempted to impose a more top-down strategy in the development of its own GFAs, trying to maintain central control rather than disperse power down to the local scale.

All GFAs need to be signed by the national affiliate union in the MNC's home country. For the Norske Skog Agreement this was Fellesforbundet, a general workers union. Subsequent research by the LO, the central trade union body, indicated that Fellesforbundet's local organizers in Norske Skog had been completely excluded from the setting up of the agreement, which had been concluded in the Brussels office of ICEM.

> Norske Skog has a Global Agreement. We looked at the Norske Skog agreement through some research in Brazil and the local unionists were a bit frustrated ... Not the fact that they didn't have the agreement ... because they love it, but they were frustrated that they were not drawn in on the consultation when they were making the Global Agreement. When I asked the shop steward in Norway, 'Why didn't you involve them [overseas trade unions]?' He said 'To be honest, I wasn't very involved either.' The project was between Mr Fred Higgs and the company with some representation from the union in Norway. But it was very much a Brussels based deal and it was not popular with the Brazilians that they were not informed when it was formalized. When I criticized ICEM on this, they said 'We couldn't ... it is difficult ... it is not something we can inform anybody about until it is finalized.' But I don't believe that. I believe it is possible to pull in more local people so that they feel ownership. They do not feel any ownership because they feel it belongs to Brussels and isn't theirs. (Interview, International Officer, LO, Oslo 2005)

This question of *local ownership* is critical to the politics of trans-national union solidarity. Deals between union bureaucrats and the chief executives of large corporations might provide good publicity for the union but does little to foster a collective consciousness among grassroots workers. Indeed, for workers who have little previous knowledge of collective action it can foster the impression that unions are just another arm of corporate management.

More positively in the Norske Skog case, a well organized local union branch in Norway set up an alternative global network of shop stewards through the establishment of a Global Works Council, much to the chagrin of the ICEM bureaucracy.

> The global works council agreement was different because that was actually worked out between the shop stewards in Norway and the company ... and Higgs came in later ... and they felt more happy about that. It was their deal and not Fred Higgs's deal ... and when I was in a meeting in Bonn last year and I met

> one of the colleagues from ICEM and I told everybody about the new agreement with Norske Skog, she was very upset, because she said, 'That's not the principle of the ICEM, we shouldn't have global agreements through works councils, we don't believe in that. So, how can you sign something like that.' So, there was a big scandal ... and everybody had to make phone calls. In the end she calmed down because the comrades from Norske Skog here said: 'Well, we will just resign. We don't want to be part of ICEM anymore if that's a problem for them.' So it was sorted out in the end. But, I find there is a problem around the process of Global Agreements. (Interview, International Officer, LO, Oslo 2005)

Ultimately, as a growing number of national affiliates are beginning to understand, for GFAs to achieve even the modest aim of monitoring global labour standards in individual MNCs and their supplier networks, local activists will need to be more empowered than at present. A lack of resources in most cases means that the national and international offices that sign GFAs are in no position to monitor or enforce them. It is only through local activists and shop-stewards and workplace-based networks that they can function effectively:

> We don't have the resources to go around to every factory in every country that is operating from Norway. We're totally dependent on our local shop stewards who are able to carry out overseas visits on company time. (Vice President, Norwegian affiliate, commenting on their recently signed GFA, interview, Oslo 2005)

To succeed, therefore, GFAs need to instill a culture of decentralization and local autonomy within GUFs. In contrast, the ICEM model involves the scaling up of a corporatist model of centralized national collective bargaining often ignorant of the diverse geographical contexts in which MNCs operate.

Putting the Global into the Local: Forging Grassroots Geographies of Solidarity

A typical response by national and international union officers to the question about grassroots involvement in building trans-national solidarity is that ordinary workers have little interest in what goes on outside their local workplace. As a national officer in one of ICEM's British affiliate unions memorably put it to us in an interview:

> Our members would say 'Why are we sending two people to work in Indonesia to unionize them when we need the two people in Wolverhampton?'. [or] 'why is he fucking about with Cuba when we've got members down in Darlington? Can't you get an officer to go down and see them? They're getting the sack in Darlington and he's buggering about in Cuba.' So you're caught in a dilemma.

They juxtapose it, they don't see it as complementary. (National Sector Secretary, British union, interview, July 2005)

A similar frustration was voiced by the international officer of a Norwegian union in getting workers at the local level to connect up their everyday struggles to more global concerns:

That is the dilemma. International work is becoming more and more important for trade union movements and I think that our union has to be involved more and more in international work than we do today but when membership is declining, the ordinary member or shop steward says, 'The issue is here and now. It's in our plant. It's the Norwegian problems with the industry. It is here that we have to use the resources and the people.' And to get the understanding of the bigger picture, and to see these things not as separate issues, I think it's quite a challenge. (Interview Vice President, Norwegian ICEM Affiliate, 16th August 2005)

While it may be true that most workers have a more pressing concern with the immediate and the local conditions confronting them in the workplace, this reflects the realities of their own class position within a global economy, rather than any necessary antagonism to international issues.

Additionally, caught up in the daily realities of working for a living, providing for families and all the time and efforts that this soaks up, particularly in a world of neoliberal capitalism where the pressures of the labour process are intensifying in many places, many people simply do not have the time or means of access to information to engage effectively in international solidarity activities. Nevertheless, it is a rather big leap in logic to write off grassroots interest in broader global issues as completely as many national and even local union officials do. Indeed, conversely, it could be argued that, given all the constraints and pressures facing people in their every day struggles to 'get by', it is surprising just how much grassroots workers do engage with transnational solidarity issues. This is also a reminder however of the geographical connections and relational solidarities that always link workers in different places to the extent that Featherstone argues against: 'the distinction between local, particularistic struggles and more universal abstract politics with global ambition' (2008: 31).

Of course, the most effective transnational campaigns are those that successfully link workers' struggles in different places to the same processes of exploitation and alienation, and with the advent of a more globalized capitalism, when these processes are bound up in the same corporation's activities it has arguably never been easier to make the connections. In our research, interviews and focus groups with shop stewards suggest a growing realization of the importance of connecting up with workers elsewhere to more effectively combat the actions of their own MNCs.

> But from our point of view, globalization is a big issue because … and companies are hiding behind the fact that … 'it's not a decision we're making locally or it's a global directive and you've gotta do it. So, we feel a loss of power in so much as what you can do locally to influence something globally. For me, that's the biggest hurdle that we certainly face.' (Focus group comment, National Shop Stewards forum, British affiliate union, October 2005.)

A more pertinent barrier to grassroots internationalism seems to be the vertical hierarchies through which ICEM operates so that grassroots workers are seldom involved in transnational union activities. From our evidence of talking to workers in the oil industry, there was a growing recognition of the need to network more at the European and global scales particularly with United States workers (Focus Group discussion, London, 2005), but they expressed frustration with existing structures such as work councils and ICEM networks.

> What kind of contact do we have internationally? I would say, it's basically zero. And, I have been a steward for about 12 years or so, I got invited once to an ICEM conference in Brussels. I know Mark has recently been to Rome for an ICEM meeting. I mean, we get all the … As branch secretary you get all the paperwork about Colombia and Iraq and whatever, but there's no other, apart from the literature, there's no other forms of interaction. (Focus group comment, National Shop Stewards forum, British affiliate union, October 2005.)

Existing forums for dialogue between grassroots workers within ICEM do little to facilitate the kind of informal exchanges that allow trust to be built and mutual affinities to develop. For example, European Works Councils tend to remain formal company-dominated forums with little room for independent worker engagement (see also Wills 2000). One UK oil worker told us of meeting an unelected women representative from southern Europe who was there 'because her boss had asked her to attend' whilst another discovered that what he thought were union representatives from Eastern Europe were actually Human Resource managers (Authors' interviews). In one extreme case, an Exxon European Works Council, meeting was held at Brussels airport on a single day with return flights booked over a restricted time period of a few hours that left little freedom at all for interaction beyond the formal timetable of meetings (Focus group comment, National Shop Stewards forum, British affiliate union, October 2005).

The more sustained and horizontal forms of trans-national solidarity between grassroots union members are, more often than not, created through the work of national affiliates, although this happens unevenly, reflecting the different geographical contexts that we have alluded to above. Trans-national solidarity is not achieved overnight and our evidence suggests that much of the more effective interactions have developed over a long time period through relations between individual unions or workers.

The British union, TGWU, has provided one of the most innovative examples of grassroots networking in recent years, through a recent joint project with the International Federation of Workers Educational Associations, funded by the UK's Department for International Development. The International Study Circles programme (WEA 2000) uses the internet to bring together groups of workers around common global themes. These have included: 'Transnational Corporations', 'Migrant Workers in the global economy', 'Women workers in the global food industry' and 'Globalization and the responses of trade unions in Asia' and have involved workers from the UK, Finland, Hungary, Peru, Philippines, South Africa, Barbados, Estonia, Bulgaria, German and Sweden.[5] Elsewhere, the Norwegian affiliates have the greatest level of resources committed to local union building activities, particularly in Africa and South East Asia. This happens through a complex spatially extensive set of relations, whereby the funding for international activities tends to come from the LO in partnership often with Norwegian civil society groups, but because the focus is always on a Norwegian MNC, projects are usually coordinated by the affiliate union with its company-based shop stewards and union representatives. In best case scenarios, such projects allow the opportunity for the build up of strong inter-personal trust relations at the local level, with funding to allow workers from Norway to spend months at a time in other countries, interacting with local workers and vice versa (WEA 2000). Norwegian unions have a long history of such bi-lateral international work and, in common with some French unions, have built up strong relationships with union organizations in former communist countries such as Vietnam and Russia, and continuing anti-capitalist regimes such as Cuba and Venezuela as a result.

Reformulating Class Politics through Space in Geographies of Transnational Solidarity

While there is a growing recognition of the importance of trans-national solidarity in the international trade union movement, there is little sign of union efforts having much impact in forging a new labour internationalism. One reason for this, as our example of ICEM attests, is the continuing set of geographical entanglements that characterize the spatial and organizational logics of the union movement, with decision-making power and resources continuing to be held at the national and international levels with a reluctance to disperse power to more decentred networked forms of organization. Theoretically, what this demonstrates is the importance of a nuanced perspective on space and place and their inter-relationship for understanding union action and the possibilities and constraints of transnational solidarity. Both territorially intensive and geographically extensive social relations need to be appreciated in the forging of new labour geographies (see Beaumont and Nicholls 2007). As we have shown, there is a critical disjuncture

5 For a report on these activities, see: www.ifwea.org (accessed 6th June 2007).

in the labour movement in the way that unevenly distributed power in the union movement across space provides certain places and their dominant actors, by dint of strong national positions, with hegemony over transnational action.

Conversely, a more positive aspect of recent labour agency is the growing confidence of some local union organizers to develop their own trans-national networks. There is arguably more potential in these networks to generate mutual solidarity because power relations are likely to be more equal, connections and communication more open and respectful of local context and difference (Holloway and Pelaez 1998, Olesen 2005). In the main however, unions continue to fail to link local and national struggles to broader politics and networks. Overcoming this situation and the promotion of transnational labour solidarity is best achieved on the one hand by empowering and including grassroots unions activists in trans-national solidarity projects, and on the other through developing a coherent and more outward-looking politics of place.

At the heart of union dilemmas in constructing more effective networks of transnational solidarity is the imperative to develop a wider spatial imaginery among grassroots workers; a broader global consciousness that is embedded within their place-based struggles over working conditions and employment rights. Drawing on the recent work of Doreen Massey (2004), this involves a more self-reflexive politics that recognizes responsibility, but also affinities, with distant others. A failing of many GUFs in this respect is their inability to develop a coherent and alternative narrative about globalization, neoliberalism and their effects that connects with the everyday realities of place-based, but not necessarily place-contained, workers. This would involve recognition of a global responsibility with *distant others* by recovering a sense of class politics across space that has infused socialist and labour internationalism from the time of Marx onwards.

Acknowledgements

The research behind this chapter was funded by ESRC project: RES 000230528, The Politics of Convergence Space in Global Justice Networks. Our thanks go to Corinne Nativel for her research assistance in this project.

References

Castree, N., Coe, N.M., Ward, K. and Samers, M. 2004. *Spaces of Work*. London: Sage.

Cumbers, A. 2004. Embedded internationalisms: building trans-national solidarity networks in the British and Norwegian trade union movements. *Antipode* 36, 829–50.

Cumbers, A., Nativel, C. and Routledge, P. 2008a. Labour agency and union positionalities in global production networks. *Journal of Economic Geography* 8(2), 369–87.

Cumbers, A., Routledge, P. and Nativel, C. 2008b. The entangled geographies of global justice networks. *Progress in Human Geography* 32(2), 183–201.

Dicken, P. 2007. *Global Shift.* London: Sage.

Featherstone, D. 2008. *Resistance, Space and Political Identities*. Oxford: Blackwell.

Gibb, E. 2005. *International Framework Agreements: Increasing the Effectiveness of Core Labour Standards*. Geneva: Global Labour Institute.

Holloway, J. and Pelaez, E. 1998. Introduction: reinventing revolution, in *Zapatista! Reinventing Revolution in Mexico*, edited by J. Holloway and E. Pelaez. London: Pluto.

ICEM 2003. *Statutes*. Brussels: International Federation of Chemical Energy and General Workers Federation.

Massey, D. 2004. Geographies of responsibility. *Geografiska Annaler B* 86(1), 5–18.

Moody, K. 1997. *Workers in a Lean World.* London: Verso.

Munck, R. 2002. *Globalisation and Labour: The New Great Transformation.* London: Zed.

Olesen, T. 2005. *International Zapatismo.* London: Zed.

Routledge, P. 2003. Convergence space: process geographies of grassroots globalisation networks. *Transactions of the Institute of British Geographers* 28, 333–49.

Routledge, P. and Cumbers, A. 2009. *The Entangled Geographies of Global Justice Networks*. Manchester: Manchester University Press.

Waterman, P. 1998. *Globalization: Social Movements and the New Labour Internationalisms*. London: Mansell.

Waterman, P. 2007. The international union merger of November 2006: top-down, eurocentric and…invisible?, in *A Political Programme for the World Social Forum? Democracy, Substance and Debate in the Bamako Appeal and the Global Justice Movement,* edited by J. Sen and M. Kumar (with P. Bond and P. Waterman). New Delhi and Durban: CACIM and Centre for Civil Society, 450–58.

Waterman, P. and Wills, J. 2001. *Place, Space and the New Labour Internationalism*. Oxford: Blackwell.

Wills, J. 2000. Great expectations: three years in the life of one EWC. *European Journal of Industrial Relations* 6, 83–105.

Chapter 5
Exploring the Grassroots' Perspective on Labour Internationalisms

Rebecca Ryland

Introduction

The emergence of neoliberalism produced a renaissance in academic debate providing reflections on labour internationalism most often vigorously advising the global working class movement to advance on an international footing: to recreate the former internationalist calls of the past, when international trade unionism flourished and multinational capitalism was in its infancy (Lee 1997). The previous cries for internationalism arose from European workers during the latter half of the nineteenth century on recognizing their commonalities through a hegemonic class consciousness, resulting in networks of solidarity aimed at countering the weakened position of their trade unions as a consequence of a dramatic increase in world trade (Munck 2002). Soon after, internationalism was seen to stagnate, relying on initiatives merely 'international in rhetoric and occasionally in practice' (Ghigliani 2005: 360) before retreating to a national, statist, protectionist stance in part as a consequence of post-1945 ideological divisions often associated with Cold War politics.

It is the apparent continuation of this stance (Pasture and Verberckmoes 1998) that has prompted many to call for trade unions to rethink the scale of their organization (see for instance Wills 1998) whilst readily associating labour internationalism with outdated, static, somewhat irrelevant policies in attempts to counteract the supposed fluid, networked, international mobility of their adversaries, multinational corporations. Positions inherent of traditional, industrial working classes; the national industrial trade union; and a traditional socialist theory are depicted as ultimately acting as a prison for internationalist consciousness and action (Waterman 2007). Mere recognition of this, as well as the processes and impacts of the neoliberal agenda, globalization and the crises of the labour movement, will not automatically lead to a new labour internationalism however (Hyman 2001). Instead, such issues force 'reconsideration of the questions of union identity and the terms of inclusion and exclusion…and supplies motives for attempting to do something, reacting collectively in an organised and responsible way' (Munck 1999: 14). Thus, many within the labour movement and academia have concluded that 'unions need new utopias, and these are unlikely to have much purchase if their focus is solely at the national level' (Hyman 2001: 175).

In articulating solidarities beyond the national level, trade unions should consider critique from within the labour movement that suggests grassroots members are often unaware of the existence of, or rationale for, labour internationalism and/or unable to make connections between the workplace and the international despite the politico-economic frameworks that make this in some ways unavoidable. It is a commonly held opinion that because of conjectural problems with economic stagnation and recession, fundamental changes in the occupational structure together with changes in management strategies, and cultural, institutional, ideological and political development trends, the working class has become fragmented and disaggregated (Hyman 1990). The capitalist agenda of fragmenting international worker consciousness into multiple, disconnected, fragmented identities may influence trade union members to perceive internationalism as unfeasible, and their trade union counterparts as the distant other for whom there exists only competition. Alternatively, a rapidly expanding global proletariat as a result of neoliberal globalization, with access to communication technologies and increased mobilities, may suggest that an effective labour internationalism could promote a reconstitution of an international working-class identity.

It is from this positioning that this chapter will attempt to frame an understanding of how trade union agendas that aim to promote labour internationalism are perceived by union members (Ryland and Sadler 2008). This is an issue that remains unexplored in any great depth with Castree (2007) arguing that Labour Geography provides facts of what a workers' group did, but no grasp of worker agency. Despite highly developed accounts of the need for a bottom-up, grassroots led internationalism (Waterman 2001), this chapter refers to empirical evidence that suggests grassroots members, notably those whose subscriptions fund international activity, often feel detached from an internationalist identity and its motivations. The chapter will support the notion that grassroots members' perspectives on labour internationalism are influenced by their 'social being' (Marx 1970) which conceptualizes a worker's wider existence, be that class/non-class identities, local affairs, global forces and moral geographies. This often enables workers to recognize a commonality of interests with other workers (Castree et al. 2004) informing *how* and *why* they make connections through international strategies. It is argued here that to achieve this commonality of interests requires explicit communications and education strategies supporting claims that international solidarity does not simply exist but requires construction.

Conceptualizing Solidarity

For many, international solidarity invokes connotations of the call in the 1848 Communist Manifesto for workers of the world to unite in socialist revolution against capitalism. Here, solidarity is constructed through affective and instrumental international activity uniting trade unionists, and workers in general, under the banner of a working-class consciousness. What should be recognized is that not all contemporary trade union members readily identify with this. Over recent decades, trade unions have witnessed a fragmentation in working-class ideology with shifts in union member attitudes from collective orientation towards individualism (Madsen 1996), reinforcing normative understandings of workers in the liberal market conception:

> ... a priced, saleable commodity being, fiercely competing against all other persons, where there is no nexus, no moral relations between persons except naked self interest. (Lambert and Webster 2007: 280)

Such understandings raise questions concerning union members' perspectives on international solidarity, most notably the tensions that can arise in its construction. Tensions can derive from different interpretations of internationalism being rooted in instrumental, economic motivations or influenced by affective ideologies. An exclusive focus on economic factors often stems from accounts on the adaptation to capitalist structures that partially characterize trade unionism (O'Brien 2005). O'Brien claims that often the assumption of this focus is that only by following the internationalizing tendencies of capitalism will trade unions adopt more internationalist tendencies (see for instance Levinson 1972). In contrast, Haworth and Ramsey (1988) and Hyman (1999) claim that a common political ideology is the basis for effective trade union internationalism. The dichotomy between economic and ideological motivations is apparent in Johns' (1998) distinction between *transformatory* and *accommodationist* solidarity. Transformatory solidarity refers to workers in one place acting to help workers elsewhere without any expectation of reciprocity from which they themselves will benefit. Alternatively, accommodationist solidarity may at first sight appear altruistic but is more about defending particular privileged (work)places in the global economy.

By contending that trade union membership is automatically conducive to altruistic solidarity with global counterparts, the beginnings of a framework are constructed that builds upon Herod's understanding of the spatial fix. The term *spatial fix* is referred to by Harvey (1982) in explaining how capital seeks to reproduce space to appropriate its own needs at particular times and locations. Herod (2001) developed this further through identification of *labor's spatial fix*, the idea that workers also attempt to produce space to meet their own requirements. From this, Herod explained that since a spatial fix is a reflection of the problem of reproduction, and since reproduction is a social and cultural as well as an economic process, it follows that different groups of workers will seek different spatial fixes that reflect their cultural and social interests (Herod 2001). What must be asked at this juncture then is how, in pursuit of international solidarity, can a potential collision between class interests or spatial praxis be averted in forging commonalities over space (Herod 2003). This issue is raised by Castree et al. (2006: 382) who frame the question:

> [A]s grassroots politics is conducted 'from' place and 'about' place, how can international solidarity then be constructed with others whose commitments, aspirations and goals are forged in quite different contexts?

Harvey (1993) posits the only way to bridge difference is to emphasize commonality. But is it too weak a logic to assume workers still exist within a definite framework of an international working class identity? And if so, can trade unions construct a common identity between workers producing support for international strategies?

For many, it is problematic to speak of a transnational union identity amongst workers where, instead, in a generalized context of changing union identities, the notion of diffuse labour identities would seem more accurate (de Sousa Santos 2007). In their genealogy of working-class internationalism, Pasture and Verberckmoes (1998) criticize Marx and Engels for identifying class as the only valid criterion of contention in constructing internationalism, whilst ignoring other variations in identity and ideology. This for Harvey (1993) has been one of the contributing factors to the weakening of working class politics: the increasing fragmentation of progressive politics around special issues and the rise of social movements focusing on identities. He claims these movements have become an alternative to traditional class politics and in some instances have exhibited downright hostility to such politics.

Amin (2003) suggests that identity politics are favoured by capitalism to achieve fragmentation within any movement deemed capable of challenging the current eco-political system. This may provide recognition that a new labour internationalism based upon promoting a class consciousness could prove effective. And yet, trade unions are working under the perception that a working class identity is dissipating, resulting in many determined to shift the debate towards articulations of internationalism that are more closely related to questions of citizenship involving broader notions of solidarity and the emergence of transnational workers' networks (Josselin and Wallace 2001). Segal (1991: 90) promotes a recognition of difference whilst uniting in commonality: 'it seems perverse to pose womens' specific interests "against" as opposed to "alongside" more traditional socialist goals'. Segal parallels Hartsock's (1990: 31) concern, calling for identification of 'the similarities that can provide the basis for differing groups to understand each other and form alliances'.

Castree et al. (2004) identify three such areas of commonality between workers. The first is their relationship to capital, which means they all need to earn a living. The second is that when this occurs, workers share a common interest over improvements in terms and conditions. The third area of commonality is that workers are interdependent through their reliance on each others' production and consumption practices. It could be assumed that this commonality of interests will foster a class consciousness especially, as stated by Hyman (1999), if workers realize the inefficiency of particularistic struggles. However, there is no guarantee that individuals will identify themselves as a class. Thus, Castree et al. (2004)

argue that inter-place solidarity needs to be actively constructed to overcome a variety of worker identities. By placing the focus of international solidarity upon issues directly relating to the local,

> the interests of one group of workers could be brought to the attention of, and supported by, workers from across space in efforts aimed at opening up the landscape and making the spatial connections between workers visible. (Herod 2003: 509)

Although Labour Geography excels in providing research into the development of transnational alliances and analyses of the impacts of specific institutionalized forms of international labour organization, the majority of debate focuses on the private sector and global corporations. Interestingly, workers in the public sector are often referred to as the least affected by direct, global competition and interdependencies when, in fact, they are becoming increasingly subject to privatization and outsourcing to globally based employers. Therefore, the privatization of government functions and the history of high union membership in the public sector makes the actions of trade unions in this sector of extra significance (Ryland and Sadler 2008). In turn, privatization raises questions as to whether local struggles should be contested within the nation state or across borders, and, therefore what is the material basis, if any, for international solidarity within this sector?

Membership Interpretations of Internationalism – A UNISON Case Study

This section explores levels of awareness of labour internationalism among public sector workers and the extent to which they recognize commonalities with other workers that may create some form of working class consciousness. The empirical data used in this discussion is drawn from research undertaken in collaboration with the North-West region of the UK public sector trade union UNISON, which has a long history of actively engaging in labour internationalism and in campaigning on issues to do with social justice in the wider field of international development. Although internationally active, the union recognizes that very little is known about membership understandings of this role.

Semi-structured interviews were conducted with a number of UNISON Branch Executive Committee members (BEC) and a sample of grassroots members was drawn upon to participate in focus group discussion. Previous engagement with grassroots members had identified that their awareness of labour internationalism was limited. Therefore, examples of affective and instrumental forms of internationalism were developed from earlier research to initiate discussions. It is recognized here that the use of such prompts imposes a framework on discussions (Maxwell 1996) and that a focus group approach has limitations, but the method has been used because it can generate authentic responses to selected themes that can be used to structure further research.

Focus group discussions and interviews began with a concentration on participants' views on the international role of a trade union, eliciting an understanding as to whether they were aware of labour internationalism and whether they support it or perceive it as a distraction from the core agenda. Initial reactions produced mixed responses. BEC members were aware of labour internationalism and its motivations. The majority were able to interpret how UNISON deployed internationalism in practice and those who were unable to provide definitive examples knew where the information for this was available, who organizes it within their branch, and they were fully supportive of why this international role occurs. In contrast, focus group participants – grassroots members – were sceptical. Many were unaware that UNISON conducts international activities, with a number of participants querying the international role, immediately associating internationalism with one-way charity: 'I think we've got enough problems over here. I want my money to stay in our country.' Instead, grassroots participants claimed the core role of a trade union is to represent its members in the workplace, depicting UNISON's role as workplace bound, with no association made with internationalism. Theirs is an economic understanding of workers; individualized and geographically rooted in their workplace and the nation state.

Both sets of participants were asked their understandings of labour internationalism and how they envisage it as employed by UNISON. During interviews BEC members provided a variety of examples of labour internationalism, without needing to be probed. International development, political solidarity, transnational alliance and internal UK internationalism were all cited, along with examples of how this is recognisable in practice by drawing connections with wider international development campaigns such as anti-racism, support for migrant workers and community cohesion. Discussions here revolved around the issue being campaigned for, as well as explanations of why UNISON made this affiliation; and any implications for UNISON members. Many made comments on the instrumental and affective motivations behind internationalism linking UNISON actions to an internationalist cause, a somewhat working class consciousness in recognition of a commonality of interests. When asked to clarify labour internationalism BEC members produced responses similar to this:

> It means the idea that *we all have the same problems to face in our workplaces in whatever country we are in* ... ideally it's where unions or union members can join together in striving to find an even starting point for all knowing unfortunately this might never happen but nevertheless trying to do what we can.

Interestingly, BEC members' responses did not imply any sense of geographical distance or a sense of the 'other' when discussing solidarity, as evident in this comment:

> I say to members 'If someone was beating up your next door neighbour would you help?' They say 'yes'. I ask them 'what if it was someone in your street, your town' etc.? They all answer 'yes'. So I say, 'ok, well what's the difference if it's someone in another country?' It's the same thing. Then they get it.

In contrast, such trade union ideals were not identified during focus group discussions with grassroots members. Instead, participants appeared disconnected from internationalist strategies and appeared unable to make a connection with foreign counterparts who were often perceived as competition, regardless of whether they were employed in geographically rooted sectors, or with UK or global employers. This understanding was also recognisable in participants' perceptions of UNISON's international activities. The following quote epitomizes comments provided within each focus group and proved a contentious issue: 'Cuba? It's a junket for the bosses'. Trade union delegations to places such as Cuba and Palestine for solidarity and information exchange were seen as an opportunity for 'the bosses to have a holiday'. In light of an apparent divergence between BEC members and grassroots members' awareness of, and interpretations of, UNISON internationalism, we must ask why this difference exists? The following sections used grassroots members' focus group discussions to throw light on this issue.

Recognizing Internationalisms

As mentioned earlier, on recognizing the limitations of grassroots members to identify and recognize the motives behind UNISON international activities, discussion was guided by the provision of examples of affective and instrumental internationalism to allow participants to identify; (1) whether they could produce examples of it in practice; (2) whether they supported each action; and (3) if they could connect with its motivations, whilst recognizing commonalities with other workers. It is to these issues that the analysis turns.

Discussion began with focus on UNISON's international development campaign for quality public services and the challenge of privatization in the UK. Initially members were quite sceptical about this raising issues of charity; issues they did not deem to be part of the role of a trade union. However, discussions identified the link between workers' needs in the workplace and the strong link with familial home life, making connection with 'social being'. They recognised their roles as UNISON members as both public service users and providers as are their international counterparts, acknowledging that as a human right everyone deserves access to quality public services. Interestingly, this was referred to in terms of a human, moral right, as opposed to signifying an explicit working-class consciousness. Discussion developed based upon the notions of human rights and democracy, drawing upon wider international development issues such as anti-racism campaigning, with participants identifying UNISON activities that seek to promote anti-racism, community cohesion and the provision of support for migrant workers. They highlighted UNISON's fight against the British National

Party [BNP] in their local communities in the run up to the June 2009 European elections. UNISON's campaign to 'Kick Racism out of Football' was cited on a number of occasions with recognition of the need to build issue based campaigns with other organizations and trade unionists. This was most evident in discussions on the recent walkouts over the use of foreign staff at a Lincolnshire oil refinery where BNP activists were seen to infiltrate pickets in attempts to incite racial hatred:

> The BNP had issues with the workforce that these companies were proposing to bring in, not the company. Now do you understand that? The company want to bring people in on less wages, want to bring people in on less rights at work and infiltrate, start the rot in this country and rather than being against the workers that they're bringing in, we should be against the company for doing this!

This participant attempted to promote a common enemy, the company employing social dumping as opposed to the other workers involved. The discussion was interesting as it raised many issues related to the current turbulent economic era in which we live, which will no doubt raise trade union members' concerns on the motives behind international solidarity based upon a working-class consciousness as well as potentially exacerbating tensions in spatial praxis.

Grassroots members were asked their views on UNISON's role beyond the UK, across Europe and internationally. They were informed that UNISON is a partner to a number of trade union and labour related organizations including the International Trade Union Confederation (ITUC), European Trade Union Confederation (ETUC) and Public Services International (PSI). The gap between individual worker perceptions and the supranational arena is quite a large ideological space as shown by the fact that members were unaware of these relations and unsure of how these partnerships are conducted. This is unfortunate given the various ways in which UK trade unions have been involved in the framing of European legislations such as the European Working Time Directive, a regulation concerned with Health and Safety requirements relevant to all UNISON workers. Issues such as these may be seen to offer a basis for solidarity with other workers and so it was queried whether grassroots members were aware of UNISON North West's links with other trade unions across borders. Members did not know of these links and so were provided with the example of the strategic partnership with Ver.di, the German public sector workers union. On querying the motives behind this, members recognized benefits of linking with unions in the same sector – a two-way exchange of information and best practice would be possible with opportunities for replication in the UK workplace:

> We do similar things locally with other local authorities within environmental health and different departments, we all share information like – I've got this problem how have you tackled it? I think if a union can do this then that's got to be a good thing.

This positive reaction appeared to be in recognition of how this would benefit them personally within the workplace as opposed to an act of solidarity, linking back to notions of the economic worker and individualism, an issue explored in more depth during this following discussion between two participants:

> I think it is important to create these types of links and solidarity in the long term because by working with other countries, in effect, for selfish reasons really, they will get better representation, better terms and conditions which will make them less options for companies to move there. So there's selfish reasons in supporting it but at the same time it's important to them too.

> I think, I look at, well I know a bit about Liverpool dockers' dispute because I've got friends down there, their most successful period was working with other people internationally because they couldn't operate a strike under the laws of this country effectively. They used other ports to strengthen their foothold and allowed them to push their point to government. Well that's an extreme example but international links are important. It's not just about fighting for somebody's human rights; but about making friends and what they can do for us in return and I don't think it's selfish.

By acknowledging that multinational companies work on an international basis, participants recognized the reasoning behind UNISON's international strategies, that there is a need to counter employers at the same level of operation. They began to develop understandings of international solidarity, identifying multinational companies as a common enemy and the nation state as having the potential to be an ineffective positioning from which to conduct struggles. From this, focus groups began to develop discussions on their international counterparts – other workers and trade unionists. One participant mentioned a magazine article produced by her employer presenting profiles of a UK and German worker comparing their daily routine. Having read this she understood:

> Workers in other countries experience the same things as here and we all want the same things really, don't we?

Although not explicitly identifying this connection as a working class consciousness, this participant understood that in learning more about her foreign counterparts she had an understanding of the commonalities they experience. In turn it could be argued that she felt more solidarity, alluding to the idea that if trade unions were to develop strategies informing grassroots members on internationalism, they may have a greater understanding of their connections with counterparts producing a potential basis for solidarity and increasing support for international initiatives.

Research did, however, reveal a tension in the apparent dichotomy between workplace issues and internationalism and the need to recognize that these issues are interdependent. Perhaps if grassroots members were more aware of this then their perceptions of internationalism could change as is evident here in a quote from a Branch Officer:

> Well, they were always saying the Cuba trip's a junket and a waste of money, so I decided to show the film to a group of low paid NHS workers employed by private contractors. They soon realized that the Cubans' protect their public services and we have a lot to learn. Before it wasn't tangible, but after the film they understood why UNISON affiliate.

The film mentioned in this interview was an educational film produced by a UNISON NW branch to highlight the plight of public sector workers in Cuba. The film aims to promote Cuba's commitment to public services and internationalism, whilst recognizing its record in education and healthcare despite the politico-economic circumstances to which it is subjected. The members who viewed the film were frustrated by the conditions they endured in their own workplace. Having been recently outsourced to private employers they were well aware of the increasing privatization of the public sector and its impact upon quality of service provision. They were negative towards the motivations behind international delegations until they identified the commonalities of experience with Cuban health workers – and yet these workers were working under far more resticted circumstances. In discussions following the screening, the Branch Officer informed me that his members spoke of how the film provided some kind of 'inspiration' with regards to what could be achieved in the UK, and recognition that delegations were not merely charitable acts but two-way exchanges of information whereupon UNISON could learn best practice from Cuba. This example provides an understanding that international solidarity must be constructed between trade union members and other workers (Hyman 1999). It does not simply exist, it requires political education and communication to promote its motives.

Concluding Remarks – Making Connections?

Branch Executive Committee members were able to identify the motives behind internationalism and recognized a number of ways in which it is employed. The reason for this may be that they receive regular information on internationalism should they choose to access it. They may not all participate on international study tours or delegations, but their geographical positioning within UNISON means that they move within networks through which such information is filtered. Their daily conversations are often internationally linked, they often work alongside and have regular contact with individuals who organize and participate in international activities, and/or they may be privy to meetings where communication in the form of videos, reports, presentations and international delegations are a regular occurrence. This complements the 'social being' of the BEC member, a person who is quite often politically conscious and able to recognize commonalities between workers.

Grassroots members' perspectives on each internationalism differed according to whether they could make a connection with the 'social being' aspect of the action. When connections were made it was because they recognized the issues, had experienced it, had received visual imaginaries or experienced empathy. This is in keeping with the idea that the 'social being' emulates a worker's wider existence and informs *how* and *why* workers make international connections. With regards to other forms of internationalism, members had either participated in an associated event, for instance 'Kick Racism out of Football', and/or could make the connection between their working life and wider existence. Irrespective of the fact that UNISON is partner to a number of global labour organizations and institutions to great benefit to their membership, participants were unable to relate to this in any way. This may be due to it being perceived as a top-down process, with which grassroots members are not involved. This merits further investigation, given that these institutions play such a large part in forming the legislative framework at workplace level, it may be that these connections could be made more explicit, perhaps then enabling members to understand the motives behind these partnerships.

The chapter argues that grassroots members' support for internationalism may be increased through recognition of the motives behind international strategies. Access to political education on the necessity for unions to be international and how the international dimension affects them in the workplace may develop understandings of the connections between all workers. Theoretically, the chapter argues for an understanding that workers make international connections according to their agency and are influenced by their social being, thus acknowledging that as theoreticians we need to look beyond workers just as 'workers' (Mitchell 2005). It has recognized that international solidarity and self-interest can be conflated. Stirling (2007), writing on trade union education on globalization, argues that members quite often join a union to meet their own needs in the workplace, but the process of political education promotes a wider picture beyond that of individual, workplace and nation based needs. This wider picture enables workers to build commonalities with other workers based upon common experiences and common discontents.

The conceptual arguments and empirical results discussed within this chapter have demonstrated the need to promote different forms of labour internationalisms and for grassroots members to be aware of their existence. It recognizes that there are many vehicles through which workers produce international connections with others. It is too simplistic to claim that workers automatically identify with internationalism and other workers according to either an idealistic working-class consciousness, or social praxis, or identity. This is not to disregard the possibility of an international solidarity motivated by a universal class consciousness, but rather, to look towards one that is constructed through recognition of common concerns directly relating to the local.

References

Amin, A. 2003. *Obsolescent Capitalism: Contemporary Politics and Global Disorder*. New York: Zed Books.

Castree, N. 2007. Labour Geography: a work in progress. *International Journal of Urban and Regional Research* 31(4), 853–62.

Castree, N., Coe, N. and Samers, M. 2004. *Spaces of Work – Global Capitalism and Geographies of Labour*. London: Sage.

Castree, N., Featherstone, D. and Herod, A. 2006. Contrapuntal Geographies: the politics of organizing across difference, in *The Handbook of Political Geography*. London: Sage.

de Sousa Santos, B. (ed.), 2007. *Another Production is Possible – Beyond the Capitalist Canon*. London: Verso.

Ghigliani, P. 2005. International trade unionism in a globalizing world: a case study of new labour internationalism. *Economic and Industrial Democracy* 26(3), 359–82.

Hartsock, N. 1990. Postmodernism and political change: issues for feminist theory. *Cultural Critique* 14, 15–33.

Harvey, D. 1982. *Limits to Capital.* Oxford: Basil Blackwell.

Harvey, D. 1993. Class relations, social justice and the politics of difference, in *Principled Positions – Postmodernism and the Rediscovery of Value,* edited by J. Squires. London: Lawrence and Wishart Limited, 85–120.

Haworth, N. and Ramsey, H. 1988. Workers of the world undermined: international capital and some dilemmas in industrial democracy, in *Trade Unions and the New Industrialisation of the Third World*, edited by R. Southall. London: Zed Press.

Herod, A. 2001. *Labor Geographies – Workers and the Landscapes of Capitalism*. New York: Guilford Press.

Herod, A. 2003. Geographies of labor internationalism. *Social Science History* 27(4), 501–23.

Hyman, R. 1990. Trade unions and the disaggregation of the working class, in *The Future of Labour Movements,* edited by M. Regini. London: Sage, 150–68.

Hyman, R. 1999. Imagined solidarities: can trade unions resist globalisation?, in *Globalisation and Labour Relations,* edited by P. Leisink. Cheltenham: Edward Elgar.

Hyman, R. 2001. *Understanding European Trade Unionism: Between Market, Class and Society*. London: Sage.

Johns, R. 1998. Bridging the gap between class and space: US worker solidarity with Guatemala. *Economic Geography* 74(3), 252–72.

Josselin, D. and Wallace, W. 2001. *Non-State Actors in World Politics*. London: Palgrave.

Lambert, R. and Webster, E. 2007. Social emancipation and the new labor internationalism: a southern perspective, in *Another Production is Possible*

– *Beyond the Capitalist Canon,* edited by B. de Sousa Santos. London: Verso, 279–320.

Lee, E. 1997. Globalisation and labour standards. *International Labour Review* 136(2), 173–90.

Levinson, C. 1972. *International Trade Unionism.* London: Allen and Unwin.

Madsen, M. 1996. Trade union democracy and individualisation: the cases of Denmark and Sweden. *Industrial Relations Journal* 27(2), 115–28.

Marx, K. 1970. *Capital,* Vol. I. London: Penguin.

Maxwell, J. 1996. *Qualitative Research Design: An Interactive Approach.* London: Sage.

Mitchell, D. 2005. Working class geographies, in *New Working-class Studies*, edited by J. Russo and S.L. Linkon. Ithaca: Cornell University Press.

Munck, R. 2002. *Globalisation and Labour: The New 'Great Transformation'*. London: Zed Books Ltd.

Munck, R. and Waterman, P. 1999. *Labour Worldwide in the Era of Globalisation – Alternative Union Models in the New World Order*. London: Macmillan Press Ltd.

O'Brien, M. 2005. Working-class internationalism, in *Marx and Other Four-Letter Words,* edited by G. Blakeley and V. Bryson. London: Pluto Press.

Pasture, P. and Verberckmoes, J. 1998. *Working-class Internationalism and the Appeal of National Identity: Historical Debates and Current Perspectives on Western Europe*. Oxford: Berg.

Ryland, R. and Sadler, D. 2008. Revitalising the trade union movement through internationalism: the grassroots perspective. *Journal of Organizational Change Management* 21(4), 471–81.

Segal, L. 1991. Whose left: socialism, feminism and the future. *New Left Review* I(185), 90–91.

Stirling, J. 2007. Globalisation and trade union education, in *Learning with Trade Unions: A Contemporary Agenda in Employment Relations,* edited by S. Shelley and M. Calveley. London: Ashgate.

Waterman, P. 2001. *Globalization, Social Movements and the New Internationalisms*. London: Continuum.

Waterman, P. 2007. Emancipating labor internationalism, in *Another Production is Possible – Beyond the Capitalist Canon,* edited by B. de Sousa Santos. London: Verso, 446–80.

Wills, J. 1998. Taking on the CosmoCorps? Experiments in transnational labor organisation. *Economic Geography* 74(2), 111–30.

Chapter 6
Navigating a Chaotic Consciousness in the Trade Union Movement

Ann Cecilie Bergene

Introduction

The question of leaders and led has a long history in both Marxist theories of mobilization and theories of trade union organization. As noted by Hyman (1975), Marxist theories on trade unions can for simplicity be divided into two camps; the optimists and the pessimists. The former argues that unions have a significant revolutionary potential, while the latter are more reserved in their assessment, with some even arguing that trade union activity might inhibit revolutionary transformation. Hyman argues that how unions have been regarded has varied over time, and even within, for instance, the writings of Marx and Engels. The issue of a particular trade union consciousness has, I would argue, been discussed on three levels. First, trade union activity and consciousness has been defined and discussed in relation to that of political parties, for instance by Lenin (1901–02). Second, the consciousness of union officials has been characterized as different to that of rank-and-file members (Luxemburg 1906, Callinicos 1995). Third, the question of the relation between leaders and led in any organization, but particularly the party, has been dealt with by among others Gramsci (Sassoon 1987). This chapter primarily discusses the latter two dimensions of trade union consciousness, and I will confine myself to the organization of workers into corporate agents in the form of trade unions and thus revisit the debate on trade union consciousness. I will let critical realism guide my approach to this theoretical discussion and will try to further develop it by introducing the concept of *chaotic consciousness*. Furthermore, I will draw on Freire and Gramsci in an attempt to theorize how to navigate this chaotic consciousness in the trade union movement. In line with Creaven (2000), I will define consciousness as the powers of abstraction, reflection and acting intentionally. Class consciousness would then mean understanding and reflecting upon the social world and conceive of vested interests in terms of the position in the class structure, and act intentionally and rationally on this basis. However, I will use the concept of consciousness in the same manner as those I am referring to unless I state otherwise. When it comes to the concept ideology it has, as pointed out by Eagleton (2007), been defined in numerous ways, some of which are incompatible. I will, in what follows, use the concept to refer to grand narratives which constitute renditions of social reality, resembling what Potter and

Wetherell (cited in Jørgensen and Phillips 1999) term 'interpretative repertoires'. My use of the term does, in other words, concur with one of the senses in which Marx and Engels used it; the value-neutral sense, 'in which ideology refers to any abstract, internally coherent system of belief or meaning' (Jost et al. 2008: 127).

Theories on the Structural Determination of the Consciousness of the Union Leadership

Agents occupy different positions in emergent and pre-existing social structures which influence their consciousness and activities through structural conditioning. This results in a situation in which different, and in many cases antagonistic, vested interests pertain to differently positioned agents (Creaven 2000). One of the structures in which agents are immersed today is the capitalist mode of production. As a structure with specific social relations, in this case relations of production, it is a determinant of the vested interests of agential collectivities as they derive from their class positioning (Creaven 2000). Conditions of life are bounded by this class positioning, as the latter shapes inequalities in access to authoritative and material resources, leaving class as one mechanism informing people's consciousness. However, the vested interests so derived are only one set among a plurality of structural determinants having a bearing on the consciousness of agents. Others are, for instance, nationality, ethnicity and gender. As relationships of exploitation, and thus of asymmetry and domination, class relations are inherently conflictual.

Callinicos (1995) argues that the union bureaucracy has material interests in limiting the scope of the class struggle to a fight for reforms. In line with the argument above, we may say that full-time union officials and workers thus, to some extent, occupy different positions in the social structure and may have slightly different vested interests. Union officials, being involved in negotiations, need to regard themselves as incumbents of intermediate structural positions and act accordingly, leading Callinicos (1995) to regard them as a distinct 'social layer'. This is because if workers and capitalists do not have opposing interests deriving from their structural positioning, then unions would not need to exist. If they do have starkly opposing interests, then compromise would not be possible or desirable for workers. So, in order to engage in negotiations, they have to think it possible to assume an intermediary position while representing only one of the sides. In other words, confining the class struggle to a fight for reforms within capitalism presupposes the view that the interests of workers and capitalists can be reconciled (Callinicos 1995). The mediatory structural position endows the union officials with a structural conditioning of their own, i.e. some differing vested interests and situational logics from the workers they represent. First of all, their consciousness will, according to Michels (quoted in Callinicos 1995), be shaped by the bourgeoisie with whom they associate. Creaven (2000: 252) agrees that it becomes a 'contradictory synthesis of the world-views and vested interests of rival or opposed classes', and 'rooted in an unstable (and normally unworkable) compromise between the interests of capital for higher profits and labour for higher wages and greater autonomy or self-determination at work'. The union leadership will thus uphold a certain politics aimed at reforming capitalism, and the union leadership is held responsible for the policy of class collaboration (Callinicos 1995). Luxemburg (1906: 87) made a similar point arguing that 'the new trade union theory' sought to 'open an illimitable vista of economic progress to the trade union struggle within the capitalist system'. Trotsky held that since unions threaten the stability and logics of capitalism, the capitalists and the state seek to incorporate them. What lends this incorporation a hand, according to Trotsky, is the corruption and the conservative ideology of the union leaders 'who from heaven knows where expect a "peaceful" miracle' (Trotsky quoted in Hyman 1975: 18). Trotsky partly answers this question by pointing out how the partial successes of unions in the past in raising the material living standards of workers strengthened the reformist outlook in the labour movement. In periods of crisis, unions are forced to make a choice between transforming themselves into revolutionary organizations or becoming 'lieutenants of capital in the intensified exploitation of the workers' since there is no longer room for reformist work (Trotsky quoted in Hyman 1975: 18). Trotsky argued that the leadership has historically chosen the second path, and reacted against every effort to resist the onslaught of capital on the part of the workers. Hence, when being incorporated, union leaders use their authority to discipline and control their rank-and-file members so that negotiations are not disrupted. In other words, incorporation is made possible by certain properties of unions and their leaders, although it would be more difficult to achieve in instances where there is less of a distance between the leaders and the rank-and-file.

Deriving these interests and politics from the structural position of union officials means that such theories regard these properties as inherent in trade unionism and renders no union, however militant, immune to this tendency (Callinicos 1995). Union officials become increasingly separated from the workers they represent as institutions of interest reconciliation are entrenched and the time spent on negotiations encourages a division of labour between full-time officials and the rank and file. According to Michels (cited in Hyman 1975), union officials are hard to replace and this gives them both the self-confidence and respect to impose their own politics on the union even in the face of opposition from workers. The officials legitimize this by referring to their experience, claiming that they know better than the workers themselves. They are able to do this partly because the members lack the information and experience needed to criticize the union.

Callinicos (1995) argues that union officials clutch on to these privileges *vis-à-vis* the rank and file. In formulating his theory of 'the iron law of oligarchy', Michels also maintained that union officials adopt a 'petty bourgeois' life-style and develop a social differentiation from the rank and file which weakens their solidarity to those they represent (see Hyman 1975). Furthermore, the security and high salaries of top union leaders have led to accusations of corruption (Lester quoted in Hyman 1975), and even arguments to the effect that they are no longer able to identify with and understand workers on the shop floor and vice versa. Since full-time union officials are removed from the shop floor to an office, the immediate conflicts arising at the workplace are eclipsed by their interest in minimizing the work load. They will thus prefer industrial relations to run smoothly most of the time. In addition, spending much of the time dealing with employers, the union officials come to see negotiation, compromise and reconciliation as the *raison d'être* of unions.

Wanting to do their tasks properly, union officials need the respect and trust of their counterpart, and militant struggles appear not only as inconveniences and disruptions, but also as a threat to their relations with employers and the stability of the union. This situation was recognized by Michels (cited in Hyman 1975), who argued that in order to be successful collective bargaining institutions, unions need to gain the goodwill, or at least the acquiescence, of both the employers and the state. Furthermore, collective agreements often involve a 'peace obligation' through which unions take on the duty of at least restraining their members from unauthorized actions, a role Hyman (1975) terms quasi-managerial. Additionally, the desire for public approval leads to moderation on the part of union officials since by expressing 'reasonable opinions' he/she will 'be sure of securing at once the praise of his [sic] opponents and (in most cases) the admiring gratitude of the crowd' (Michels quoted in Hyman 1975: 16).

Callinicos (1995) argues that ensuring the efficient running of the organization might become an end in itself, and that this might even come into conflict with the active fight for improvements in wages and working conditions. Similarly, Luxemburg (1906: 86) observed that the establishment of a union bureaucracy in Germany meant:

> the over-valuation of the organisation, which from a means has gradually been changed into an end in itself, a precious thing, to which the interests of the struggles should be subordinated. From this also comes that openly admitted need for peace which shrinks from great risks and presumed dangers to the stability of the trade-unions, and further, the overvaluation of the trade-union method of struggle itself, its prospects and its successes.

Callinicos (1995: 22) argues that the union leadership is influenced by 'a sense of collective responsibility which makes them reluctant to rock the boat'. Moreover, in order to win negotiations they have to think tactically around issues which would not arise without negotiations. When, for instance, unions do need to engage in industrial action, they may do so more for the tactical reason of flexing their muscles in order to strengthen their bargaining position than to insist on the extreme demands they flag during the action. In addition to showing strength, industrial action will also have the function of convincing management that they are responsible agents doing what is in their power to restrain excessive aggression from the workers. Therefore, once, in their eyes, a satisfactory result ensues from the negotiations, officials will see no point in further agitation (Mann 1973).

Since even the struggle for minor improvements in wages and working conditions may threaten the stability of the system, the union leadership will time and again intervene in order to prevent these struggles from getting out of hand, and often call them to a close on terms falling short of the initial demands from the workers (Callinicos 1995).

Creaven (2000) employs the term bureaucratic sclerosis for a situation in which corporate agents, when finding themselves involved in social struggles or actions, try to constitutionalize, or even repress, the militancy of the primary agents they represent and obstruct any radical structural change which could be in the members' interests. According to Mann (1973), this contradiction is nowhere as clear as in the case of socialist unions, since the situational logics pertaining to unions engaging in negotiations often clash with their socialist ideology. This is in line with Michels' (cited in Hyman 1975) contention that if a union is to be effective as such, it will need to pursue policies that are acceptable to the capitalists and the state, regardless of what the political objectives of the union are. Hence, while the class understanding of union officials and members in socialist unions tend to be acute, this does not necessarily have any bearing on the actions of socialist unions, and they are therefore not necessarily fully class conscious. Mann (1973: 37) argues that despite their ultimate goals, they are still 'implicated in the day-to-day bargaining process which keep greased the wheels of capitalism'. Hence, Callinicos (1995) maintains that while it is common among leftists to argue that the main division in the union movement is ideological, the fundamental division is the conflict between the bureaucracy and the rank and file. He thus argues that the centripetal forces binding all union officials together are stronger than the centrifugal forces pushing them apart, although he admits that there are deep political divisions among union leaders. In contrast, Dan Gallin (personal communication 23.04.08), former General Secretary of the IUF[1] and currently chair of the Global Labour Institute, argues that ideology is a major determinant of how union officials understand their work and how they undertake it, and he thus disagrees with those maintaining that the union leadership necessarily assumes a certain mentality. To the extent that structural positions affect the work of unions, he argued that the fact that they are representing agential interests deriving from the capitalist social structure makes them more anchored than, for instance, political parties. Although many political parties, such as labour parties, have their origin in the trade union movement, they have lately drifted away from working-class politics. This might be because party members and leaders are no longer recruited from the trade union movement and/or along class lines, and therefore that parties no longer aspire to be mass parties. According to Moody (1997), this drift and the subsequent tension between unions and labour parties has led to a politicization of unions, especially as they confront neoliberal government policies.

1 The International Union of Food, Agricultural, Hotel, Restaurant, Catering, Tobacco and Allied Workers' Associations.

Chaotic Consciousness

Starting from the critical realist position on the distinction between abstract and concrete, I will use the concept of 'chaotic conception' to inform my theorization of consciousness. While the abstract is defined as a one-sided focus and as an imagined closed system, the concrete is 'the chaotic resultant of a plurality of generative mechanisms' derived from a multitude of underlying and unobservable structures at the level of the real (Creaven 2000: 13). In other words, class consciousness is abstract in the sense of being an analytic filter that only lets class relations through. However, in any given situation the concrete understandings and conceptions of an individual might be the result of multiple determinants, i.e. chaotic. To complete the analogy, people live in concrete situations and their understandings and conceptions are thus often derived from experiences and impressions passed simultaneously through multiple filters before reaching the eye. Hence, how situations and events are conceptualized, understood and acted upon will be determined by a number of different ideological influences and interests in 'the muddy waters of the concrete'. I will thus introduce the concept of *chaotic consciousness* in an effort to better grasp this multiple determination. Chaotic consciousness should be defined as abstractions, reflections and intentional actions that result from a plethora of ideological influences and interests originating in concrete situations, and not abstract understandings of the rules of *one* game, i.e. the abstract capitalist one, in isolation. The obverse of chaotic consciousness will thus not be true or authentic consciousness, but abstract and logical consciousness, which is one-sided and consistent. Abstract class consciousness can thus be defined as the extent to which people one-sidedly and clearly understand their existence in terms of class relations and act accordingly.

Most of the abovementioned perspectives on unions rest on one-sided structural explanations. However, as we have seen, individuals live their lives in concrete social formations, not in abstract modes of production, and most primary agents' thoughts and feelings are concrete in the sense of being the result of multiple determinants. Society cannot be reduced to one social structure, for instance capitalism, and there are thus numerous structural positions defining vested interests beside that of class, such as ideological persuasion, gender, age and ethnicity. Although Marxists would argue that class is primary and that the other structural positions map onto it in a way that facilitates 'divide and rule', the fact that people conceive their situation and interests along the lines of their positioning in these other structures ensures their causal efficacy in determining intentional actions.

Spontaneity or Direction?

Hyman (1975) argues, rightly in my view, that the classical theories on trade unions, if rendered unqualified, are overtly simplistic and may be regarded as distortions. For instance, unions can neither be regarded as solely in opposition to capitalism nor as merely a component of it. Rather, Hyman argues, there is a dialectical relationship between them and capitalist society. On the one hand, unions would not exist if there was no conflict of interest between workers and capitalists, and their establishment ensures that agential interests deriving from workers' structural position within capitalism come to the fore, thus producing an awareness of a separate identity as against capital. This awareness might in turn develop into class consciousness as workers start to reflect and act intentionally along class lines, a line of thought facilitated by the unions. Hyman (1975) thus suggests that the logic and conflicts underlying the formation of unions persist and act as counter-tendencies to the mechanisms of integration, oligarchy and incorporation identified by Lenin, Michels and Trotsky respectively. On the other hand, as we have seen above, in their daily activities unions tend to take capitalist society for granted and adjust their policies and work accordingly, in some instances to the detriment of any long-term transformative objectives. Furthermore, Hyman argues, in line with Trotsky, that the dialectic between opposition and acceptance is tied to the ups and downs of the capitalist economy and to the extent that organized workers accept the concessions capitalists are able and willing to give based on the distribution of power between the two agential collectivities and the available margin. However, this interpretation rests on an understanding of class consciousness as deriving solely from the objective conditions under which workers work and live. I would argue that consciousness derives neither solely from objective conditions or lived experience nor solely from theoretical principles from which aspirations and praxis can be deduced. Rather, consciousness derives from a more abductive reasoning in which people move between ideologies, considerations of their own interests, situations, explanatory persuasiveness and previous experience.

Callinicos (1995) reaches the conclusion that the rank and file should become less dependent on full-time union officials. This leads us to the discussion about the relation between spontaneity and direction. Primary agents may react spontaneously and explode 'into radical unrest or protest, often outflanking those corporate agents (in terms of militancy) which reputedly represent their interests' (Creaven 2000: 169). Primary agents, despite their lack of collective organization and an articulated strategy, 'often embody a keen political awareness of their own agential interests, and with this a high degree of anger and resentment at the manner of which society and its vested interests operate to deny or frustrate these' (Creaven 2000: 170). However, primary agents need corporate agents oriented to their vested interests if their demands are to be fully formulated and effective in achieving structural change, and collective action presupposes some form of common interests. While Lenin (1901–02) regarded spontaneity in solely negative terms as unconscious and instinctive action, Gramsci (1919–20) was of the opinion that the two need to be balanced. In an effort to arrive at the appropriate relationship between factory councils and unions, he agreed that the development of the latter entailed concentration and generalization of power and discipline in a strong central office. On the other hand, he does agree with Lenin that it is important to keep a certain distance from the fickle spontaneity and naïve ambitions of the masses. Furthermore, unions have the power to impose an industrial legality in the dealings between workers and their employers through their ability to negotiate agreements and take on responsibilities. Although the introduction of industrial legality involves a risk of being incorporated, Gramsci regards it as a great victory for the working class, warning instead that unions should not rest on their laurels since it is a compromise and thus neither permanent nor an ultimate or definitive victory. Still, Gramsci (1919–20) sees the potential pitfalls of trade unionism, and proposed a strategic relationship between them and workplace organizations in order to forestall, for instance, the overvaluation of the organization. While unions prevent both the capricious impulse of the rank and file from leading to set-backs or defeats and any outright class conflict, workplace organizations are more spontaneously radical. However, as pointed out by both Moody (1997) and Wills (2001) economic insecurity and the fear of job loss is strong among the rank and file, and this might act as a conservative force and a major barrier to developing trust and solidarity between different workplaces. This was to some extent recognized by Gramsci (1919–20), who notes that it is essential for the ruling class to undermine organizations external to the factory and territorially centralized, such as unions.

Gramsci regarded the division between leaders and led as the fundamental problem of politics (Sassoon 1987). He thus provided a critique of bureaucratic centralism:

> The workers feel that the complex of 'their' organization, the trade union, has become such an enormous apparatus that it now obeys laws internal to its structure and its complicated functions, but foreign to the masses [...] [B]ureaucracy sterilizes the creative spirit [...] These *de facto* conditions irritate the workers, but as individuals they are powerless to change them: the worlds and desires of each single man are too small in comparison to the iron laws inherent in the bureaucratic structure of the trade-union apparatus. (Gramsci 1919–20: 9–10)

Gramsci also argued against a situation in which party members are only connected to the leadership through a generic loyalty and where these members are there 'simply to be manoeuvred without a role of their own' (Sassoon 1987: 167). According to Creaven (2000), the British trade union movement is a dismal example of how corporate agents experience political degeneration and paralysis when the primary agents they represent lack the confidence needed to spur their representatives into action in defence of their interests.

Although Gramsci's discussion of bureaucratic versus democratic centralism originally concerned the organization of the political party, I will here extend his insights to unions. This can be substantiated both on the grounds that Michels (cited in Hyman 1975) argued that unions were even more prone to oligarchy than other political organizations, as well as what my informants have told me about the operation of unions to the effect that democratic centralism might be attainable.

First, Hyman (1975) argues that Michels' theory of the iron law of oligarchy does not pay sufficient attention to countervailing tendencies, such as pressures towards democratic practice and operating at scales other than the national. Hyman thus points out that rank-and-file activists, who interact regularly with union officials, exercise strong influence on them in the direction of democracy. Shop-floor unionism may thus represent a crucial link between the rank-and-file members and the union leadership, and through shop stewards and delegates workers may influence union policies and actions, or even choose to act independently (Hyman 1975). However, shop-floor representatives are not immune to the efforts at incorporation recognized by Trotsky, and Hyman points out how in Britain there have been several attempts to neutralize shop stewards by formalizing their role through integration into both official trade union structures and institutions of collective bargaining.

Turning to Gramsci's democratic centralism, this is defined by a leadership (of the party) which actively relates to a changing reality:

> a 'centralism' in movement – i.e. a continual adaption of the organisation to the real movement, a matching of thrusts from below with orders from above, a continuous insertion of elements thrown up from the depths of the rank and file into the solid framework of the leadership apparatus which ensures continuity and the regular accumulation of experience. (Sassoon 1987: 162–163)

He thus conceived of a dialectical relationship between spontaneity and direction in which the most important feature of the leadership is their ability to look at the broader picture and to discern what is similar in seeming diversity. Furthermore, Gramsci argues that this identification and organization of commonality ought to be done in a practical and 'inductive' way, that is, closely tied to the processes among the rank and file, as opposed to deductively and in an abstract way. Freire and Shor (1987) argue that while intellectuals first and foremost describe and use concepts, the masses of the people start by describing concrete reality. However, I would argue that organizing and identifying interests purely in an inductive way from concrete situations more easily lends itself to chaotic consciousness and losing sight of long-term goals. I have therefore argued that consciousness is best advanced by abductive reasoning.

In order to avoid the pitfalls discussed above, Gramsci urged the union leadership to develop the 'capacity to react against force of habit, against the tendency to become mummified and anachronistic' (Sassoon 1987: 164). Bureaucracy is a main obstacle to achieving this, as it is, according to Gramsci, dangerously conservative, and might, as we have seen, end up feeling that it stands on its own and is independent from the rank and file. In other words, bureaucratic organizations encourage the idea that they transcend their individual members, an attitude Gramsci labels 'fetishistic', the effect of which is the paralysis recognized in the quote from Gramsci above when members feel that the organization exists despite them and not because of them. A way of countering this tendency is for the leadership to gain the ability to analyse concrete situations, and tailor strategies and tactics accordingly. Furthermore, they need to actively involve the rank and file as 'intellectuals' in discussions, ensuring that strategies and tactics are based on their experience. Here I think the trade union movement has a lot to learn from Freire's 'Dialogical Method' of teaching which is counterpoised to any vanguardist understanding of the relation between leaders and led. Not only does communication confirm or challenge the social relations between people engaging in it, it is also an act of remaking knowledge for all participants (Freire and Shor 1987). That union officials and union members blame each other for not being militant enough, and for lacking the 'right' consciousness and/or solidarity testifies to a lack of communication and dialogue in knowing reality and setting goals. Knowledge, and certainly consciousness as it has been defined here, is '*not* an exclusive possession of *one* of the subjects doing the knowing, one of the people in the dialogue' but rather 'mediates the [...] cognitive subjects [and is] put on the table *between* the [...] subjects of knowing' (Freire and Shor 1987: 99, emphasis in original). Although direction is inherent in all education, critical direction can be ensured through letting people with training pick and present the objects of knowledge, since they have experience in producing knowledge and knowing what has been produced. Despite this previous experience and knowledge of the object under discussion, these 'educators' have not exhausted all possible ways of knowing and understanding the phenomenon, and they thus relearn through the conversation and their ability to know the object is remade. The union officials may thus have the theoretical understanding and the bird's-eye view necessary to take on directive responsibility, but need to discover with members and other workers the topics most pressing for them. Furthermore, direction should also entail re-examining topics which were previously taken for granted through acquiring a certain distance from it by abstracting it from the familiar, concrete situation. According to Freire and Shor (1987), critical consciousness means a systematic understanding of impressions, and constitutes an effort to overcome naïve understandings of reality. It is arrived at through starting from the concrete, from common sense, and through dialogue, moving to more abstract and rigorous interpretations. Concrete consciousness is regarded as inconsistent, and Freire and Shor (1987: 112) argue that it is the task of leaders, in their case teachers, to pick up on these contradictions and to 'design a pedagogy that inserts itself into the tangle'. The Dialogical Method of teaching thus involves uncovering key topics and finding access points to consciousness in order to recompose them into critical investigation asking 'the led' to re-perceive any prior understandings.

Gramsci (cited in Sassoon 1987) concludes that the answer to the question of democracy versus the 'iron law of oligarchy' rests on what he terms the class difference between leaders and led. If such a class division exists within the organization, which Gramsci argues pertains to unions and social democratic parties, then the issue becomes political. If, however, it does not exist, the division between leaders and led is merely a technical division of labour in which, through 'apprenticeship', followers are recruited to leadership. To forestall a development of the leadership into a bureaucracy, Gramsci maintains that there should be a numerous intermediate stratum between the leaders and the led.

Concluding Remarks

People find themselves in different positions in emergent social structures which influences their consciousness and activities. This happens through a structural conditioning that results in a situation in which different, and in most cases antagonistic, vested interests pertain to agents differently positioned in the structures (Creaven 2000). Consciousness was defined as the powers of abstraction, reflection and acting intentionally. Class consciousness would then mean understanding and reflecting upon the social world and conceive of vested interests in terms of the position in the class structure, and acting intentionally and rationally on this basis. Individuals belong to concrete social formations, and most primary agents' thoughts and feelings are concrete in the sense of being the result of multiple determinants, among them ideologies. Class consciousness, the extent to which class positions inform understandings and actions, being no exception, would also at any given point be what I termed *chaotic* as there are numerous structural positions defining vested interests beside that of class. Although Marxists would argue that class should have primacy, and that the other structural positions map onto it in a way that might facilitate 'divide and rule', the fact that situations and interests are conceived of along the lines of these other structural positions ensures their causal efficacy in determining intentional actions. In line with the above discussion, it would be wrong to regard what has been termed a trade union consciousness as consciousness. Rather, union officials seem to have a chaotic consciousness, i.e. they appear to draw on several, even contradictory, ideologies and/or interpretations in their understanding of the capital-labour relationship. Furthermore, their structural positioning vests them with interests which make them liable to let the ideology of social partnership and reconciliation inform their understanding and hence consciousness. However, we have also seen that the notion of a shared understanding among union officials is an oversimplification since they are embedded in a variety of structures which will have a bearing on how they abstract, reflect and decide upon actions. The division between rank and file and leadership or bureaucracy is too rigid and there is often no clear-cut distinction between the perspectives offered by union officials, delegates and workers, although there may be tendencies in the direction of the union leadership employing, or being held responsible for, a less militant, and a more narrow and reformist, class-collaborationist approach.

References

Callinicos, A. 1995. *Socialists in the Trade Unions*. London: Bookmarks.

Creaven, S. 2000. *Marxism and Realism. A Materialistic Application of Realism in the Social Sciences*. London: Routledge.

Eagleton, T. 2007. *Ideology: An Introduction*. New and Updated Edition. London: Verso.

Freire, P. and Shor, I. 1987. *A Pedagogy for Liberation. Dialogues on Transforming Education*. South Hadley, Massachusetts: Bergin & Garvey.

Gramsci, A. 1919–20. *Soviets in Italy*. Nottingham: The Institute for Workers' Control, 1974.

Hyman, R. 1975. *Marxism and the Sociology of Trade Unionism*. London: Pluto Press.

Jørgensen, M.W. and Phillips, L. 1999. Diskursanalyse some teori og metode. Fredriksberg: Roskilde Universitetsforlag.

Jost, J.T., Nosek, A. and Gosling, D. 2008. Ideology: its resurgence in social, personality, and political psychology. *Perspectives on Psychological Science* 3(2), 126–36.

Lenin, V.I. 1901–02. What is to be done?, in *On Trade Unions. A Collection of Articles and Speeches*. Amsterdam: Fredonia Books, 2002.

Luxemburg, R. 1906. *The Mass Strike*. London: Bookmarks Publications, 2005.

Mann, M. 1973. *Consciousness and Action among the Western Working Class*. London: The Macmillan Press Ltd.

Moody, K. 1997. *Workers in a Lean World: Unions in the International Economy*. London: Verso.

Sassoon, A.S. 1987. *Gramsci's Politics*. Second Edition. Minneapolis: University of Minnesota Press.

Wills, J. 2001. Uneven geographies of capital and labour: the lessons of European Works Councils. *Antipode* 33(3), 484–509.

Chapter 7
Schumpeterian Unionism and 'High-Road' Dreams in Toronto's Hospitality Sector

Steven Tufts

Introduction

North American labour unions continue to struggle against neoliberal capital and states (Panitch and Swartz 2003, Fantasia and Voss 2004). Much normative discussion has focused on the change in structures and strategies necessary for organized labour to renew itself and increase union power. In the US, the formation of the *Change to Win* (CTW) coalition of unions which broke away from the AFL-CIO in 2005 was, on the surface at least, a response to significant political and strategic differences over how to organize workers (see Fletcher Jr and Gapasin 2008 for a critique of CTW). The innovative multi-scalar strategies developed by UNITEHERE, a CTW union formed through an uneasy merger of textile (UNITE) and hotel (HERE) workers in 2004, to organize the hotel workers in Canada and the United States have been widely cited as exemplars of union renewal (Tufts 2007, Schenk 2005, Gray 2004, Wills 2002, Moody 2007). Yet, this renewed hotel workers' union has a past and present rife with contradiction. For example, the union has a long history of affiliation with organized crime throughout North America and was under US Supreme Court Monitorship until the late 1990s with many corrupt locals under trusteeship. More recently, the union continues to support Republican candidates in the US, especially in gaming states, and large scale development projects such as the Olympics which provide jobs for union members as they displace the poor (Tufts 2004).

These persistent contradictions limit positioning UNITEHERE as a 'social justice', 'social movement' or even 'social' union. At the same time, the recent period of renewal is a significant break from past 'business union' practices. This chapter proposes 'Schumpeterian unionism' as a theoretical framework capable of characterizing the complex and contradictory practices of labour unions such as UNITEHERE. Schumpeterian unions do not fall into the business, social or social movement unions categories proposed by Kumar and Schenk (2006). Instead, they attempt to take an active role in the 'creative destruction' of their sectors and own organizational practices. Schumpeterian unions are constantly in a process of rescaling their activities in order to regulate evolving capital formations. The proposed framework is developed through a juxtaposition of two ideal types of unionism. Defensive Atlantic unionism, a model characterizing many business unions following the post-war compromise, is juxtaposed against the model of ideal social movement unionism which is found in much of the literature on labour union renewal since the 1990s.

The theoretical discussion is, however, also grounded in the recent experience of Local 75 UNITEHERE, the largest hotel workers union in Toronto representing over 7,000 workers in a city where approximately 75 per cent of large hotels are unionized. Local 75's efforts to build 'high-road' partnerships in Toronto's hospitality industry and organize the sector are the empirical basis for the theoretical model. The chapter begins with a brief account of Local 75's recent project to organize a poor community around 'high road' labour market development initiatives. This narrative is followed by a theoretical discussion of Schumpeterian unionism, a conceptual framework for understanding specific union practices. Developed are the model's intra and extra institutional relationships as well as their different relations with capital and the state.

Toward High-Road Partnerships in the Hospitality Sector

In March 2003, a global outbreak of Severe Acute Respiratory Syndrome (SARS) spread to a number of countries from its origin in southern China's Guangdong Province. In Canada, the outbreak was largely isolated in Toronto and resulted in 44 deaths in the city leading the World Health Organization to issue a travel advisory for the city which devastated its tourism industry and displaced thousands of hospitality workers at the start of the tourism season (Tufts 2003). In response to the situation, Local 75 with the assistance of the labour friendly Labour Education Centre (LEC), established a Hospitality Workers Resource Centre (HWRC) with government support to assist the workers with employment insurance, job searches and retraining for work within and outside the sector (Tufts 2009). Despite the closure of HWRC following the SARS crisis, Local 75 continued to support models to explore ways to reinvent the low wage sector as a high wage productive industry with training and advancement opportunities.

Leading up to the 2006 round of collective bargaining the local union formed a task force (in which the author participated) which produced a report, *An Industry at the Crossroads: A High Road Economic Vision for Toronto Hotels.* The report advocated 'high road' partnerships between capital and the state with aims to: strengthen internal labour markets by developing well defined career ladders, upgrade the skills of all workers, provide training for entry level workers, develop the Toronto tourism industry, and address the short and long-term labour requirements of the sector. The recommendations in *Industry at the Crossroads* called for higher wages and benefits, greater union representation, equal opportunity in the workplace, new training capacities (i.e. a union administered training facility similar to the Culinary Training Academy in Las Vegas), as well as social investments in housing, day-care and transit to make the city 'liveable' for low wage workers. In part, the document was inspired by a report issued by the Working for America Institute (2003) on hotel work.

In key collective agreements negotiated with large hotels in 2006, an Equal Opportunity Training Fund (EOTF) was established to provide resources for worker training. As the fund was embedded in a collective agreement, management is now obliged to support and participate in joint labour-management training initiatives which were previously voluntary. The first stages of a new training centre are now underway. In a unique deal with the City of Toronto and the Toronto Community Housing, an 85 unit affordable housing coop is being built in the heart of downtown with some units marked for hotel workers and 2,400 square feet for a new hospitality training centre on the first floor. As Paul Clifford, Local 75 president, explained in an announcement of the fund's birth in an agreement with the Fairmont Royal York:

> We learned during SARS, that so many workers were frustrated in not being able to afford to upgrade their skills and move up. Finally we've achieved the means to do this with this Training Fund. (UNITEHERE 2005)

Taking the High Road to the Community: Real Jobs for Rexdale

Rexdale, in northwest Toronto, is a community struggling with underemployment and poverty where many of Local 75's members live. It is also the location of the Woodbine Racetrack, the city's horse racing and slots facility owned by a private non-profit firm, the Woodbine Entertainment Group. The racetrack, established in the 1870s, is on the 266 hectare Woodbine lands, the largest track of undeveloped land in Toronto. The land, purchased long ago by the city's 'horsemen' is also part of an 'employment district', land designated by the City of Toronto to be protected from residential and retail development given the scarcity of industrial land.

In 2005, the Baltimore based Cordish Company announced a massive 'urban revitalization' project in Rexdale based on an expansion on the existing racetrack facilities. The initial $310 million investment was reported to generate 2,300 permanent jobs and $150 million in taxes per year for the first decade (Cordish 2005).[1] More recently, the employment generated is estimated at 10,000 jobs and the private investment is now claimed to be over $1 billion. The project entitled 'Woodbine Live!' mimics other projects of Cordish such as the 'Power Plant Live' urban revitalization project in Baltimore's inner harbour.

The financial success of the Cordish Company is largely based on significant tax incentives from the local state for its development projects in communities challenged by disinvestment. It is an unlikely coincidence that Woodbine Live! emerged at a time when new tax incentive programs became permissible in Toronto after the City of the Toronto Act was passed in 2006 by the provincial government granting the city new taxation powers. Toronto's first significant tax incentive

1 All monetary figures are in Canadian dollars.

program was the development of Tax Increment Equivalent Grants (TIEGs). TIEGs give firms multi-year tax holidays if an investment falls in a targeted region or sector. In the case of Toronto, these are being aimed at specific sectors and investments in employment districts. The economic sectors targeted for support by the City of Toronto include screen based industries such as film and television, aerospace, pharmaceuticals, and/or electronic equipment manufacturing, food and beverage manufacturing, environmental production and research, IT and new media, life science industries and research, and tourism, which make the Woodbine Live! investment eligible. Any qualified additional or new investment in a district and/or specified sector would receive a 100 per cent municipal tax holiday in the first year and the taxes would be 'rehabilitated' in annual increments for a number of years. In the case of Woodbine Live! the proposed TIEG is over 20 years and would save Cordish over $120 million in municipal taxes. There are fundamental criticisms of such development schemes as they largely involve local states picking winners (neighbourhoods and/or industries) to receive the tax break and there is always a possibility the firm will exit the market quickly after receiving the largest tax breaks at the front end of the incentive. Further, rounds of inter- and intra-regional competition are set off as every region and sector competes to qualify (LeRoy 2008).

Given the size of the development project, the accommodation, food service and retail sector components, and the fact that many of Local 75's members live in Rexdale, the union launched a community campaign in 2006 to intervene in the development process and hopefully secure and Community Benefits Agreement (CBA) with Cordish. The Community Organizing for Responsible Development, using the acronym 'CORD', as a direct affront to Cordish, organized numerous community meetings in order to educate the community about the sometimes brutal impacts of such developments on communities, e.g. increase in housing prices, decay of the industrial tax base, and the proliferation of precarious low wage service employment. The community campaign led by Local 75 insisted that the city negotiate a CBA with Cordish which would include guarantees of 'high road' economic (e.g. local hiring and training funds), social (e.g. recreation centres and mixed income housing), and environmental (e.g. public transit access and green space) benefits in exchange for the large tax incentives (Tufts 2008). The strategy was largely drawn from the experiences of US cities, where communities have entered such agreements with developers as a means of securing 'good jobs' (Gross 2005).

CORD had some success in mobilizing the community to question the benefits of the retail and entertainment development which would have generated largely low wage service jobs. Large turnouts (over 500 people) to CORD events calling for 'Real Jobs for Rexdale' emphasized the need for local hiring targets and training funds from the developer (Monsebratten 2007a). Cordish initially expressed a willingness to talk about demands for 30 per cent local hiring (Monsebratten 2007b), but in interviews with union activists, the company initially 'walked away' from any such concessions. At present, the only concession the Cordish Company has made, despite the lucrative TIEG, is with respect to modest local hiring and some training support in the community – much less than CORD's demands.

While CORD has not abandoned its campaign to pressure the developer, it has had relatively little success in exerting pressure on the company through the local state which was overwhelmingly supportive of Woodbine Live![2] It is too early to provide a definitive analysis as to why the campaign achieved relatively little, but there are some points which can be made with respect to Local 75's largely 'top-down' model of community unionism and coalition building. First, the primary organizers of the campaign are directly employed by Local 75. The omnipresence of UNITE-HERE paid staff and activists at CORD events and local hearings brings into question the extent to which 'community' is actually involved in the campaign. Some reduced CORD to a union's attempt to leverage a neutrality agreement with the developer in order to organize any new hotel properties associated with the project. Second, after some initial discussion over lobbying to expand the qualifying criteria for tax incentives to include things such as neutrality provisions, it appears that the strategy by local labour shifted toward having some input into the list of which sectors provide 'good jobs' for Toronto's new economy. There was less criticism over TIEGs as a development strategy and the path they create toward unbridled interregional competition.

Lastly, it was suggested by activists that there were conflicts within UNITE-HERE over tactics and at which *scale* to pressure the developer. The Cordish Company is an international development firm with projects in several North American cities. While CORD was pressuring the company in Toronto, it was reported by informants that the international union leadership was also attempting to secure a comprehensive continental community benefit framework from the company. As a result, international leadership based in Washington, local leadership in Toronto and community activists in Rexdale were not always working toward the same political objectives. Of particular debate was the extent to which the community coalition should denounce the project as key demands were left unaddressed. Here, we have the classic geographical dilemma facing unions as the interests of one organizational scale (e.g. an international union in a larger framework with Cordish) conflict with those of another (e.g. a local union fighting one project) (Castree et al. 2004).

Theoretically conceptualizing the contradictions surrounding Local 75's community organizing efforts within dualistic categories such as 'business' and or 'social unionism' is problematic. Commentators have recognized the need to transcend false binaries characterizing discussions of union renewal such as servicing versus organizing models since their inception (Fletcher Jr and Hurd 1998). I propose below Schumpeterian unionism as a conceptual framework for a

2 The author deputed against the TIEG proposed for the project at the City of Toronto's Economic Development and Planning and Growth Management Committees. The tax incentive had broad based support of staff and elected officials (City of Toronto 2007).

model of unionism which is embedded in deeper theoretical accounts of evolving labour-capital-state relations.

Schumpeterian Unionism

In a groundbreaking article, Jessop (1993) forwarded the Schumpeterian workfare state (SWS) as model of state-capital relations displacing the Keynesian Welfare State established in post-War Atlantic Fordist economies. At its core, the SWS model characterized a number of national policies aimed at implementing the neoliberal project, e.g. labour market flexibility, innovation. The model is steeped in regulation theory and inspired by Schumpeter's (1942: 83) treatise on economic evolution that centred the process of 'Creative Destruction' as 'the essential fact about capitalism' that must be understood in order to understand overall economic development. Over a decade, Jessop (2002) refined his initial model to where he speaks of Schumpeterian Workfare Post-national Regimes (SWPRs) as the successor to the Keynesian state. The SWPR's focus is on: economic policies which increase competitiveness in global markets, putting downward pressure on social wages with limited welfare, the rise of networks of public-private governance, and rescaled state policy above and below the nation. From a geographical viewpoint, the most significant evolution of Jessop's model is the integration of how capitalist states re-scale economic policy to global (e.g. policies allowing capital to flow to low wage regions) and sub-national levels (e.g. policies enhancing regional metropolitan competitiveness rather than national economic development). Indeed, it was the initial aspatiality of Jessop's model that inspired a number of geographers to explore how restructuring of the welfare state was being played out across space and differentiated in places (Peck 1996, 2001, Peck and Jones 1994). Brenner (2004) has also built upon Jessop's work to define how the reterritorialization of state policies from the national-global and national-local has created a number of contradictory new state spaces. In particular, national policy supporting cities and the decentralization of power has shifted the governance and reconfigured neoliberal state-capital formations toward the urban (Brenner and Theodore 2002).

For labour unions, the implications of these new state spaces must not be understated as they shape the quantity and quality of jobs delivered by capital and the ability of workers to organize. In most cases organized labour has been the target of SWPRs, but it is problematic to theorize labour as outside processes shaping variations of capitalist states. While many emergent labour union renewal strategies can still be discussed as a reaction to neoliberal restructuring and the regulatory environment imposed by SWPRs, other emerging union structures and practices can also be viewed as an integral part of contemporary capitalist economies. In other words, it is consistent to consider how unions are implicated in various formations of neoliberal regimes since Schumpeterian economies require the consent of Schumpeterian unions, which is under-theorized by Jessop (2002).

Table 7.1 **Schumpeterian Unionism**

Union Activity		Defensive Atlantic Unionism	Ideal-Renewed Social Movement Unionism	Schumpeterian Unionism
Intra-institutional organizing	*Recruitment*	Blitzing worksites	Intensive and broad-based bargaining	Strategic campaigns, selective targets
	Servicing and collective bargaining	Staff servicing, concessionary bargaining	Membership involvement, mutual aid, participatory bargaining	Servicing efficiencies, bargaining innovation
Extra-institutional organizing	*Coalition building*	Union paternalism	Social movement unionism	Strategic alliances
	International solidarity	Symbolic	Global Unions	Situated networks
	Mergers	Survival, raiding	General Unionism	Strategic capacities, opportunistic
	Central labour bodies	Irrelevant	Formative	Re-imagined, multi-scalar
Labour-management Relations	*Labour-management cooperation*	Concessionary	Workplace militancy	'tactical cooperation', minimal trust
	Training	Employee responsibility, minimizing skills requirements	New training opportunities, lifelong learning, class-based education	Vocational competitive training, administrative union education
Labour-State relations	*Economic development*	Coercive, flexibility	Social democratic partnership	Ephemeral, Multi-scalar, uneven tripartism
	Labour market regulation	Exclusive	Broad-based, centred at the margins	Sectoral, management of dissent

To elaborate the model of Schumpeterian unionism, I juxtapose the model against two antithetical ideal-types: defensive Atlantic unionism, associated with the rise and decline of Atlantic Fordism and renewed social movement unionism (see Table 7.1). For each ideal type I compare four elements of union activity: intra-institutional organizing, extra-institutional organizing, labour-management relations, and labour state relations.

Intra-Institutional Organizing

A renewed social movement unionism calls for more intensive organizing strategies, e.g. house calls to new members, that build solidarity for long term struggle against capital rather than immediate increases in recruitment (Bronfenbrenner 2003). An ideal renewed unionism is also not an abandonment of servicing members. Instead, such renewed unionism calls for new approaches to servicing that allow workers to self-organize against an employer through mutual aid (Bacharach et al. 2001) and shift any surplus resources created by greater membership participation to organizing new members into a broad labour movement beyond traditional representation, e.g. representing workers by pressuring employers prior to any formal union certification.

Schumpeterian unions are traditional in the sense that they do not adopt structures advocated by commentators who argue that low-wage service workers require entirely new frameworks for bargaining across entire occupations and communities rather than by the single worksite (see for instance Clawson 2003, Wial 1993). Instead, organizing campaigns are strategic and specific targets are identified. Servicing by paid staff is not abandoned, but collective bargaining efficiencies are developed and linked to organizing. For example, collective bargaining, such as gaining neutrality agreements with employers, is used to exert demands for greater employer recognition of unions and to facilitate organizing. Most important, campaigns and bargaining are increasingly multi-scalar as local initiatives are linked to global struggles in complex ways (Tufts 2007, Sadler and Fagan 2004). Mechanisms creating multiple scales of organizing are, however, rarely painless transitions and are contested within and among unions as power is (de)centralized in organizations to match rescaled capital (Savage 2006). In other words, Schumpeterian unions destroy the geographical scale of defensive unionism by shifting strategic and other organizing resources to new sectors and spaces. In the case of Real Jobs for Rexdale, CORD was very much a top-down campaign orchestrated very much from the local union office and the specific strategies designed by paid organizers to fight a multinational development firm. Tension did arise, however, between local efforts and the international office attempting to engage CORD at the international scale.

Extra-Institutional Organizing

In terms of organizing outside the union, defensive Atlantic unions largely retreat to insular strategies where coalitions are rare and strictly managed, international solidarity is symbolic at best, or at worst a means of disciplining communist unions in other countries, mergers and raids are carried out for survival, and national and regional central labour bodies are largely irrelevant. In contrast, building power over employers is often directly linked to union engagement with other labour and non-labour organizations in order to build a broad 'social movement unionism' around diverse communities. Again, such broad-based multi-issue coalitions are rarely achieved. And while 'Global Unions' are deemed crucial for organizing large-scale neoliberal transnational employers such as Wal-Mart (Bronfenbrenner 2007), transferring resources to national or international central labour bodies and coalitions which can negotiate and enforce effective global agreements with employers remains a challenge (Stevis and Boswell 2007).

Schumpeterian unionism approaches extra-local organizing in a much more flexible and contingent manner. First, all unions exercise caution when entering coalitions. It is better to characterize many of these relationships as ephemeral coalitions rather than sustainable community unions as they are campaign specific and labour's superior financial and political power is not easily surrendered to create equal coalitions (Tatersall 2005).

Social movement unionism works toward establishing powerful global unions rather than symbolic bureaucratic lobbying bodies, but a healthy scepticism must be levelled at the new labour internationalism (Waterman and Wills 2001, Munck 2002, Castree et al. 2004). Nevertheless, workers are forming a range of new relationships that transcend national borders. It is often more accurate, however, to describe the emerging formations as situated networks where actors exchange information that can be put into action at a number of different scales rather than formal institutions able to leverage power against transnational capital (Wills 2002).

Schumpeterian unions also re-imagine the role of central labour bodies when they can contribute to multi-scalar campaigns and practices. Cooperative organizing campaigns, the mediation of jurisdictional disputes, and the economies of scale gained through joint lobbying and educational efforts are part of this re-imagined but still limited role for central labour bodies.

In fragmented labour movements with many small unions, mergers are viewed as a means of rationalizing necessary resources. While defensive unions enter mergers for survival (Chaison 1996), a renewed social unionism views mergers as a fundamental reorientation toward a general unionism reminiscent of One Big Union. Schumpeterian unionism use mergers to increase their strategic capacity to organize in new sectors (Yates and Ewer 1997).[3]

3 In the case of UNITEHERE's merger in the summer of 2004, the significant financial resources of one union in a declining sector (i.e. UNITE) combined with a poor union established in a growing sector (i.e. HERE) to increase the capacity to organize workers in immigrant communities. The merger has not been as effective, however, and

For Schumpeterian unions, extra-local organizing is crucial, but such alliances are most often ephemeral and driven by the practical requirements and the specific scale of its struggle with employers. Theoretically grounding Schumpeterian unionism in the case of CORD, accentuates this point. Very much a 'top-down' coalition, Local 75 engaged the community for very limited strategic purposes (i.e. to leverage power over the developer). While this was not an insular strategy it is far removed from the depth of coalitions theorized by others (Tattersall 2005). Further, the strategies themselves were imported through networks of activists and experiences with good jobs coalitions internationally.

Labour-Management Relations

Clearly, unions in a defensive position with capital and the state have engaged in concessionary relationships with employers. Lower wages and increased labour flexibility in terms of job security and the production process are the central pillars of capital's challenge to the regulated labour market (Vosko and Stanford 2004). In order to secure institutional survival and the jobs of members unions have actively participated in the workplace restructuring and made concessions to remain competitive. In many cases, concession bargaining has been the norm as workers struggle to maintain jobs in industries facing significant competitive pressures. Such concessions are counter to social movement unionism's commitment to a significant reassertion of workplace democracy where workers have a significant control in the implementation of new technologies and work practices, the distribution of work, the management of working time, and the development of life-long learning and class-based education for all workers (Livingston and Sawchuck 2003).

Schumpeterian unions do continue to cooperate with employers on a daily basis, but the relationship is better characterized as one of 'limited trust' where cooperation is often ephemeral and lent for competitive support under specific conditions. Some unions have adopted strategies which engage labour-management partnerships as new trade offs are made between increasing productivity and maintaining job security. 'Social partnerships' are argued to strengthen labour's position by 'trapping' capital investment in local markets through 'high road' strategies that emphasize training and increased productivity in the workplace (Kelly 2004). It is the 'high-road' model of labour market investment which is advocated by Schumpeterian unions rather than class-based education and is evident in the emphasis CORD and Local 75 placed on training and standards for Woodbine Live! rather than as denouncing the project as an inappropriate form of economic development.

---

by 2008 cracks between the two 'halves' threatened the future of UNITEHERE as a single union.

Labour-State Relations

Cooperation with the state was a hallmark of defensive Atlantic unionism. Economic development strategies often found labour and the state in a tight tripartite relationship with local capital to attract international investment through civic boosterism (Harvey 1989, Hudson and Sadler 1986). Union leadership was also implicated in the coercive management of dissent against neoliberal states as they restructured social wages in order to attract investment. In the worst cases, labour acted as imperialist agents for transnational capital as they disciplined labour and expanded into the global South.

An ideal renewed social movement unionism emphasizes greater social democratic control over investment decisions which will better manage industrial (over)capacity and the creative destruction process. Further, social movement unionism defines union membership in broad terms which are centred on the needs of workers in the margins rather than exclusive formal membership. For example, the Living-Wage Campaigns which emerged in the 1990s in the de-industrialized US, pressured the state and employers to improve working standards for all workers in specified low-wage sectors, not only unionized workers (Luce 2004).

Schumpeterian unionism also enters uneven tri-partite relationships. There are however, two main differences between its approach to economic development and more defensive unionism. First, there is recognition of the unevenness of the partnerships and for this reason they are often temporary initiatives based on specific issues rather than a long-term cooperative framework. Second, Schumpeterian unions are engaged in a process of king out and exploiting new state spaces created by the re-territorialization of states in an increasingly global economy (Brenner 2004). Several changes in the re-scaling of advanced capitalist economies have placed large urban centres at the heart of accumulation strategies (Jonas 1999, Brenner and Theodore 2002). Cities have increasingly become central to regional and national economic development strategies as post-industrial engines of growth based innovation and technology, centres of concentrated capital and elite labour exerting greater territorial control over production and distribution networks, and cultural leaders producing dominant representations of the city spaces of global media. The rise in the concentration of power and wealth in the metropolis has been accompanied with income polarization and the demand from local governments for greater political and economic power to address such issues.

The Stronger City of Toronto for a Stronger Ontario Act, more commonly referred to as the *City of Toronto Act*, passed by the Ontario legislature in 2006 did give a broad range of powers to the city which included the ability to establish new revenue streams and tax incentives to attract industry and employment to depressed areas. Local 75 has attempted to re-territorialize their relationship with the state to the local level as they look identify the new points of leverage created by re-scaled state accumulation strategies. While Schumpeterian unions remain concerned with national electoral politics (UNITEHERE spent significant resources on electing Barack Obama in 2008), they increasingly focus on finding ways to exploit the slippages created when states download power and responsibilities to urban scales. Securing local state investment in human capital to enhance competitiveness and productivity, i.e. high road initiatives, in a few sectors which are difficult to outsource is one strategy. In this respect, Schumpeterian unions situate themselves inside the process of 'creative destruction' rather than as profound agents of economic transformation. There is less emphasis on broad policy reform and benefits are secured only for workers fortunate enough to belong to unions in specific industries which can elicit minimal government support for their sector, e.g. subsidies for training and tax subsidies for employers.

Theorizing New Labour Geographies

Recent commentary on the status of Labour Geography has called upon researchers to export geographical theory to mainstream labour studies and industrial relations (Castree 2007, Herod et al. 2007). This chapter has integrated some geographical concepts into a broader theoretical framework for understanding labour union renewal in a changing economy.

Schumpeterian unionism is only one attempt to link union action to changing advanced capitalist economies. Juxtaposing two ideal types of unionism against one another to locate a model of 'actually existing' unionism is important if discussions of labour union renewal are to move beyond the normative. Asking 'what is to be done' will remain crucial, but understanding how and why unions are responding in specific, and at times contradictory, ways is important. Schumpeterian unions are not transformational and at best will be able to 'hold the line' against neo-liberalism given some minor state regulatory reforms. Identifying the limitations of existing approaches is a first step in making the case for bolder experimentation.

Furthermore, this chapter has attempted to use political economic theory as a means of understanding changing union strategy, but the theory is also very much grounded in real cases of labour and community action. Grounded theoretical approaches are a huge strength of contemporary economic geography (Yeung 1997). New theoretical work must remain grounded in real political circumstances of workers' lives and Labour Geographers are well positioned to contribute in this way. Abstracting labour too close or too far from the realities of everyday political struggles leads to fatalist or unrealistic analyses of what is politically achievable in neoliberal spaces.

References

Bacharach, S.B., Bamberger, P.A., and Sonnenstuhl, W.J. 2001. *Mutual Aid and Union Renewal: Cycles of Logics of Action*. Ithaca and London: Cornell University Press, ILR Press.

Brenner, N. 2004. *New State Spaces – Urban Governance and the Rescaling of Statehood*. London: Oxford University Press.

Brenner, N. and Theodore, N. 2002. Cities and the geographies of 'actually existing neoliberalism', *Antipode* 34(3), 356–86.

Bronfenbrenner, K. 2003. The American labour movement and the resurgence of organizing, in *Trade Unions in Renewal: A Comparative Study,* edited by P. Fairbrother, and C. Yates. London: Continuum, 32–50.

Bronfenbrenner, K. (ed.), 2007. *Global Unions: Challenging Transnational Capital Through Cross-Border Campaigns*. Ithaca, New York: ILR Press and Cornell University.

Castree, N. 2007. Labour Geography: a work in progress. *International Journal of Urban and Regional Research* 31(4), 853–62.

Castree, N., Coe, N., Ward, K. and Samers, M. 2004. *Spaces of Work: Global Capitalism and the Geographies of Labour*. London: Sage Publications.

Chaison, G. 1996. *Union Mergers in Hard Times*. Ithaca, New York: ILR Press.

City of Toronto, 2007 Staff Report. *Stimulating Economic Growth: Toronto's Approach to Financial Incentives.* City of Toronto, 21 November 2007.

Clawson, D. 2003. *The Next Upsurge: Labor and the New Social Movements*. Ithaca, New York: Cornell University.

Cordish Company 2005. *Development Plans Announced for Woodbine Live! $310 Million Entertainment District Planned*, press release, 28 July 2005.

Fantasia, R. and Voss, K. 2004. *Hard Work: Remaking the American Labor Movement*. Berkeley: University of California Press.

Fletcher Jr., B. and Gapasin, F. 2008. *Solidarity Divided: The Crisis in Organized Labor and a New Path Toward Social Justice*. Berkeley: University of California Press.

Fletcher Jr., B. and Hurd, R. 1998. Beyond the organizing model: the transformation process in local unions, in *Organizing to Win: New Research on Union Strategies*, edited by K. Bronfenbrenner, S. Friedman, R.W. Hurd, R.A. Oswald, and R.L. Seeber. Ithaca, New York: Cornell University Press, 37–53.

Gray, M. 2004. The social construction of the service sector: institutional structures and labour market outcomes. *Geoforum* 35, 24–34.

Gross, J. 2005. *Community Benefits Agreements: Making Development Projects Accountable*. Washington: Good Jobs First and the California Partnership for Working Families. Available at: www.goodjobsfirst.org.

Harvey, D. 1989. *The Condition of Postmodernity*. Oxford: Basil Blackwell.

Herod, A., Rainnie, A. and McGrath-Champ, S. 2007. Working space: why incorporating the geographical is central to theorizing work and employment practices. *Work, Employment and Society* 21(2), 247–64.

Hudson, R. and Sadler, D. 1986. Contesting work closures in western Europe's industrial regions: defending place or betraying class?, in *Production Work, Territory*, edited by A. Scott and M. Storper. London: Allen and Unwin, 172–93.

Jessop, B. 1993. Towards a Schumpeterian workfare state? Preliminary remarks on post-Fordist political economy. *Studies in Political Economy*, 40, 7–39.

Jessop, B. 2002. *The Future of the Capitalist State*. Cambridge: Polity Press.

Jonas, A. (ed.), 1999. *The Urban Growth Machine: Critical Perspectives Two Decades Later*. Albany, New York: State University of New York.

Kelly, J. 2004. Social partnership agreements in Britain: labor cooperation and compliance. *Industrial Relations* 43(1), 267–92.

Kumar, P. and Schenk, C. 2006. Union renewal and organizational change: a review of the literature, in *Paths to Union Renewal: Canadian Experiences*, edited by P. Kumar and C. Schenk. Peterborough, Ontario: Garamond, Braodview and CCPA, 29–60.

LeRoy, G. 2008. TIF, Greenfields, and sprawl: how an incentive created to alleviate slums has come to subsidize upscale malls and new urbanist developments. *Planning & Environmental Law* 60(2), 3–11.

Livingstone, D. and Sawchuk, P. 2003. *Hidden Knowledge: Organized Labour in the Information Age*. Lanham, MD: Rowman & Littlefield Publishers.

Luce, S. 2004. *Fighting for a Living Wage*. Ithaca, NY: Cornell University Press.

Monsebratten, L. 2007a. Woodbine Live deal galvanizes community. *Toronto Star*, 10 May, R1, R4.

Monsebratten, L. 2007b. Racing to corral Woodbine jobs. *Toronto Star*, 26 June, A10.

Moody, K. 2007. *US Labor in Trouble and Transition: The Failure of Reform from Above, the Promise of Revival from Below*. New York: Verso.

Munck, R. 2002. *Globalization and Labour: The Great New Transformation*. London: Zed Books.

Panitch, L. and Schwartz, D. 2003. *From Consent to Coercion: The Assault and Trade Union Freedoms*. 3rd Edition. Aurora, Ontario: Garamond.

Peck, J. 1996. *Work-place: The Social Regulation of Labor Markets*. London: Guilford Press.

Peck, J. 2001. *Workfare States*. New York: Guilford Press.

Peck, J. and Jones, M. 1994. Training and enterprise councils: Schumpeterian workfare state or what? *Environment and Planning A* 27, 1361–96.

Sadler, D. and Fagan, B. 2004. Australian trade unions and the politics of scale: reconstructing the spatiality of industrial relations. *Economic Geography* 80(1), 23–43.

Savage, L. 2006. Justice for Janitors: scales of organizing and representing workers. *Antipode* 38, 645–66.

Schenk, C. 2005. Union renewal and precarious employment: a case study of hotel workers, in *Precarious Employment: Understanding Labour Market Insecurity in Canada*, edited by L. Vosko. Montreal: McGill-Queen's University Press, 335–52.

Schumpeter, J. 1942(1976). *Capitalism, Socialism and Democracy.* 5th Edition. London: Allen and Unwin.

Stevis, D. and Boswell, T. 2007. International framework agreements: opportunities and challenges for global unionism, in *Global Unions: Challenging Transnational Capital Through Cross-Border Campaigns*, edited by K. Bronfenbrenner. Ithaca, New York: ILR Press, 174–94.

Tattersall, A. 2005. There is power in coalition: a framework for assessing how and when union-community coalitions are effective and enhance union power. *Labour & Industry* 16(2), 97–112.

Tufts, S. 2003. SARS and new normals: healthcare and hospitality workers fight back. *Our Times: Canada's Independent Labour Magazine* 22(4), 16–21.

Tufts, S. 2004. Building the 'competitive city': labour and Toronto's bid to host the Olympic Games. *Geoforum* 35(1), 47–58.

Tufts, S. 2007. Emerging labour strategies in Toronto's hotel sector: toward a spatial circuit of union renewal. *Environment and Planning A* 39(10), 2383–2404.

Tufts, S. 2008. Labour and (post)industrial policy in Toronto. *Relay: A Socialist Project Review* 23, 30–33.

Tufts, S. 2009. Hospitality unionism and labour market adjustment: toward Schumpeterian Unionism? *Geoforum* 40(6), 980–90.

UNITEHERE 2005. *Hotel Workers Union and Fairmont Royal York Settle Contract Talks*. UNITEHERE press release, 17 Nov. Available at: www.unitehere.org/presscenter.

Vosko, L. and Stanford, J. (eds), 2004. *Challenging the Market: The Struggle to Regulate Work and Income*. Montreal: McGill-Queen`s.

Waterman, P. and Wills, J. (eds), 2001. *Place, Space and the New Labour Internationalisms*. Oxford: Blackwell.

Wial, H. 1993. The emerging organizational structure of unionism in low wage services. *Rutger's Law Review* 45(3), 671–738.

Wills, J. 2002. Bargaining for the space to organise in the global economy: a review of the Accor-IUF trade union rights agreement. *Review of International Political Economy* 9(4), 675–700.

Working for American Institute 2003. *US Hotels and Workers: Room for Improvement.* Washington: WAI.

Yates, C. and Ewer, P. 1997. 'Changing strategic capacities': union amalgamations in Canada and Australia, in *The Future of Trade Unions: International Perspectives on Emerging Union Structures,* edited by M. Sverke. Aldershot: Ashgate, 131–48.

Yeung, H. 1997. Critical realism and realist research in human geography: a method or a philosophy in search of a method? *Progress in Human Geography* 21(1), 51–74.

Chapter 8

Trade Unions as Learning Organizations: The Challenge of Attracting Temporary Staff

Dorit Meyer and Martina Fuchs

Introduction

Similar to other organizations, unions must learn how to deal with changing socio-economic conditions in order to safeguard their survival. Recently, unions in Germany have been facing two problems, both an increase in temporary staff which the unions cannot recruit through traditional shop-floor strategies and a general loss of members in the last two decades. This has spurred unions to seek out potential new members that were ignored until now. Among these are temporary blue-collar and white-collar workers, hereafter referred to as temps.

Employing the concept of *dynamic capabilities*, we will investigate how trade unions are trying to solve both of the abovementioned problems; they are recruiting temps to increase their membership. In this chapter we demonstrate that the learning process of such organizations is not all-encompassing but multi-locational and multi-scalar. Multi-locational means that an organizational entity, here a local union, in one region learns in a different way from a local union somewhere else. Multi-scalar means that the learning processes occur at different hierarchical levels and entails spatial diffusion. The multi-scalar learning processes are played out both from the bottom up and from the top down at different hierarchical levels in large organizations. Our central argument is that multi-locational and multi-scalar processes are not decisive in the development of dynamic capabilities but often aid in the necessary adaptation of the organization to changing socio-economic conditions.

In this chapter we will turn our attention to the concept of dynamic capabilities before we will differentiate between multi-locational and multi-scalar capabilities. After providing some brief remarks on methodology, we will delve into the current changes in socio-economic conditions and consider the necessity for unions in general to adjust. From there we will point out the extent to which dynamic capabilities of unions differ regionally and at different organizational levels.

Dynamic Capabilities

The Concept of Dynamic Capabilities

In organizational sciences and in economic geography, the interdisciplinary discussion on learning processes within organizations refers in general to large-scale enterprises, not to trade unions. Indeed, there are differences between big companies and unions: unions are organized primarily according to principles of elections, appointments, acceptance and confirmation, and are thus characterized by being more 'bottom-up'. Companies, on the other hand, follow the principles of the market economy (Drinkuth et al. 2003). Nevertheless, unions, as well as companies, must continually adapt to new conditions and live up to the expectations of their members or customers (Herod 1998, 2003, Lier 2007, Peck 1996). Hence, it appears justified to transfer the analysis of capabilities, which was originally intended for commercial companies, to trade unions.

Dealing with organizations in general, we need to go back to the resource-based view. Originally, the resource-based view was an instrument for identifying the strategic resources of a firm. In this perspective, the resources of an organization, i.e. labour, capital and environmental conditions, play an important role. The basic idea is that the competitive advantages of a company lie in its bundle of valuable resources. If such resources are heterogeneous and not perfectly mobile, they can convert a short-run competitive advantage into a sustainable one. If these conditions prevail, the bundle of resources can assist the company in sustaining above-average returns (Argyris and Schön 1978). In the present context of increasing competition for knowledge in high-tech firms, the resource-based view highlights the importance of VRIN-type assets which are 'valuable, rare, inimitable, non-substitutable competences' (Schreyögg and Kliesch 2006: 456).

Still, the resource-based view has a significant shortcoming: It highlights *assets*, i.e. what competences the firm is currently in possession of. Yet, VRIN assets by themselves do not explain the competitiveness of a company sufficiently. This is because some low-tech firms without VRIN assets or with very few VRIN assets have a quite strong and sustainable competitive position in the global economy, and because there are companies which possess VRIN competences but fail anyway.

Consequently, this view tends to focus on specific assets and has a tendency to ignore what organizations can *do*. Therefore, the term 'capabilities' (competences or skills) fits better than 'knowledge'. Capabilities are derived from an epistemology of practice rather than an epistemology of possession. Organizations can expand their capabilities especially by developing and applying *new routines*. Routines can here be understood as a solidification of practices and as standardized procedures. They are thus patterns of behaviour and interaction that represent successful solutions to a particular problem such as the problem of attracting temps. They can be informal rules of conduct, or they can be formalized in institutions, as, in our case, in tariff agreements.

The term 'capability' implies that socio-economic conditions do not exert any decisive influences on the internal processes of gaining competences. The reason for this is that agency plays an important role in addition to external structural conditions. This is in line with the recent emphasis in economic geography on the argument that internal processes and agents play their part in controlling the strategy of an organization, and not simply a set of location factors. When the organization is a high-tech company, expensive technologies and key patents are considered focal points and understood as resources of knowledge. When the organization is a trade union we can think of high quality human resources and their capabilities to react innovatively to problems.

From the perspective of organizational sciences, dynamic capabilities are necessary for an organization. The dynamic capabilities approach is a further development of the abovementioned resource-based view. Dynamic capabilities have some specific attributes. Firstly, they are usually defined as the ability of an organization to integrate, extend, enlarge, build and reconfigure internal and external competences. Secondly, they are an innovative reaction to a rapidly changing environment (Eisenhardt and Martin 2000, Teece and Pisano 1994). Dynamic capabilities allow organizations to react to complex socio-economic dynamics and to solve difficult and complex tasks. Hence, dynamic capabilities do not merely solve everyday or standardized problems (Schreyögg and Kliesch 2006). In contrast to the resource-based view that focuses on assets as resources that the company possesses such as knowledge, dynamic capabilities are seen as changing routines. For companies, this implies integrating external routines, such as external technology or customer behaviour, and integrating and reconfiguring it inside the organization. In the case of trade unions, this means that they understand the characteristics of temps and their interaction with trade unions, and that they integrate, extend, enlarge, build and reconfigure such knowledge in such a manner that the entire organization can find a new strategy to organize temps.

Dynamic capabilities generate various new formal and informal routines. When the organization finds a new strategy, the routines become part of a superior, complex new logic defining how the organization behaves in general. The complex pattern and arrangement of routines that have evolved in an organization is referred to as the 'architecture'. The architecture also includes a 'construction plan', the blueprint, prototype or model, to further processes of learning (Amin and Cohendet 2004). The architecture is reproducible, and influential actors in the organization want the routines to be reproduced. However, in multi-locational and multi-scalar organizations, dynamic capabilities may differ from location to location and between the different hierarchical levels. Although the concept of dynamic capabilities is quite wide-spread in organizational sciences, how spatially differentiated organizations are able to generate dynamic capabilities at its different locations and its different levels has not been at the centre of attention. Competences are spatially and hierarchically dispersed as organizations themselves are inherently spatial (Massey 1984, Fuchs 2008).

Dynamic Capabilities from a Multi-locational and Multi-scalar View

Large organizations are not homogenous entities, but are internally differentiated (see Dodgson 1993). Processes of learning vary between the different locations of an organization and between the various hierarchical levels, which also have specific spatial extensions. This is true for large multi-plant companies as well as for other organizations, such as trade unions.

The following analytical approaches, involving three different levels of analysis, may further the understanding of the development of dynamic capabilities with regard to the unionization of temps. Firstly, the local, regional and national structures of the labour market and local initiatives reacting to the problems may impel organizations to develop dynamic capabilities. In other words, the socio-economic environment may be such that it strongly influences unions to recruit temps. We refer to this as *the structural approach*. Secondly, practices of trade unions vary between regions and organizational levels, and recruitment of temps may not necessarily be the outcome of changes in the socio-economic environment. The outcome will reflect formalized and informal relations between trade unions, works councils and managements. Works councils are organizations representing employees. Their role is to be the intra-firm counterpart to management. At the national level there are equivalent structures of relationship between trade unions (e.g. IG Metall) and national employers' associations. Works councils exist in a range of institutional forms in the EU, including Germany (Betriebsrat), Spain (Comité de empresa), and France (Comité d'Entreprise). In Germany, works council representatives are also appointed to the Board of Directors (Aufsichtsrat). These are actors who are embedded in their specific trade union culture (Castree et al. 2004, Jonas 1996), which in turn suggests that strategies *vis-à-vis* temps differ spatially. We refer to this approach as *the organizational approach*. Thirdly, when acting in a beneficial organizational context, individual actors might drive forward processes of learning and generate dynamic capabilities. In this case, it is not the pressing problems of the socio-economic environment that lead to dynamic capabilities, nor the more passive embeddedness in a specific trade union culture, but active agents anchored in the organization who act in an advantageous context. We refer to this approach as *the person-oriented approach*.

Methodology

Since little is known about dynamic capabilities in trade unions, our method is qualitative-explorative. A qualitative approach is needed in order to uncover the reasons, the *modus operandi* and the conditions in which the routines are developed and adopted. Such a qualitative design also helps to access sensitive information.

The study was conducted by a country-wide inquiry in 2008. Based on an analysis of the internet homepages of various IG Metall administrative units, potential interviewees responsible for organizing temps were identified. Thirty-seven trade union secretaries at local and district administration offices in all parts of Germany were subject to semi-structured interviews, and three interviews were made with members of the executive board of IG Metall. In addition, each interviewee was asked to recommend further union secretaries to be interviewed, employing the snowball method. All interviewees were working with new strategies for recruiting temps and have contributed substantially to the organizational learning process. Textual data has been analysed in a systematic manner by qualitative content analysis which offers a synthesis of openness and theory-guided investigation.

The Development of Dynamic Capabilities in Unions

In the following we will show that the suitability of the structural approach, organizational approach and person-oriented approach differs according to spatial scales and organizational levels. The structural approach might explain the general strategy of the executive board of trade unions. However, to understand multi-locational and multi-scalar dynamic capabilities, the person-oriented approach seems more adequate. Obviously, individual actors can only exert influence when their actions are corroborated by colleagues. Hence, the organizational approach which concentrates on the network relations in and between organizations highlights the embeddedness and connectivity between agents implementing new routines.

The Structural Approach: Changes in the General Strategy of the Trade Union's Executive Board

Over the last 30 years, unions have been losing members in the core economies (Dølvik and Waddington 2004, Phelan 2007). In Germany this development started gradually in the 1980s. After an increase from 7.97 million trade union members in 1990 to 11.80 million in 1991 that resulted from the reunification of Germany, there was a dramatic decline to 6.34 million members in 2009 (DGB 2009).

One of the most important reasons for the general loss of members is to be found in the industrial transition. With the decreasing significance of industrial production, traditionally a trade union stronghold, unionization declined sharply (Behrens et al. 2003). The situation was made considerably worse by the extensive loss of jobs in the new federal states. Furthermore, there was an increase of companies without tariff agreements.

In addition, the use of temps increased enormously during the same period. Temps are not directly employed by the company but through a temporary employment agency and may be both blue-collar and white-collar workers. Employment agencies assign the workers to the company in question, and temps are seldom union members (Dörre et al. 2006). Exact figures are not available but estimates in 2003 suggest that the degree of union organization of temps was less than 5 per cent, and thus far below the degree of union organization of regular private sector employees which amounted to approximately 17 per cent (Vitols 2003). Yet, a gradual increase in the number of temps from about 22,000 in 1983 to 385,000 in 2004 has been documented (Federal Employment Office 2008, 2009). This growth accelerated due to changes in the temporary employment legislation as part of the general reform of the German labour market law ('Hartz IV-reform'). As part of the legal restructurings, many former stipulations regulating temporary work were removed. For this reason the number of temps soared to 800,000 in July 2008. This increase in the number of temps has become an important factor in the decrease of union members in Germany since the reunification.

Unionizing temps might offer a possibility for reversing this trend. Thus, adhering to the structural approach, we can state that the increase in non-unionized labour on the one hand and the increase of temps on the other hand impelled the executive boards of unions to develop new strategies. In the course of the 1990s, the search for a new understanding of trade union cultures began. Consequently, new routines evolved and were established. The acceptance of temporary employment on the part of the DGB unions manifested itself in 2003 in the tariff agreements with the three temp agencies and thus as formal routines. This preceded the changes in temporary employment legislation. In addition to the removal of many restrictions, pay and treatment of temps comparable to permanent staff became compulsory as a result of the legal restructurings. In order to circumvent the principle of equal pay and equal treatment, agents in the temporary employment branch pressed for finalization of tariff agreements. When the unions succumbed to demands for negotiations on the part of temp agencies, the unions thereby officially accepted temporary employment as a method of reducing unemployment. After the tariff agreements of the DGB unions in 2003, a discreet bottom-up process of dispersal of new routines *vis-à-vis* temps began.

However, the employment conditions of temps still differ fundamentally from that of the dominant group of union members (Druker and Stanworth 2004). Temporary workers represent a sector that has so far not been covered by formal industrial relations regulation in Germany. The works councils are required by law to cater for permanent employees, but not the temps. Furthermore, the German unions had to change their own way of thinking. Until recently, they termed temporary work 'modern slavery' and claimed that this form of employment should be banned. Nowadays, the unions have begun to realize that their rejection of the booming sector cannot remain forever. The loss of members and the increase of temps persuaded the German Trade Union Federation (DGB) and their affiliates to change their attitude. Their executive boards now had the task of diffusing and distributing the new strategy to the different locations and branches of their organization.

The Person-oriented Approach and the Organizational Approach: Multi-locational and Multi-scalar Perspectives

Unions have a decentralized organizational structure with administration offices at the local level and district administrations at the larger regional (sub-national) level. Agents are, in our case, union officials and secretaries in particular. They are waged union employees responsible for the political and operational business of the union. Important impulses emanate bottom-up from agents in administrative offices to neighbouring and sometimes even distant administrative offices or districts of the union.

Strategies directed at recruiting temps were developed in concert with works councils at temp agencies and at firms employing temps, and through interaction with temps themselves (Martin and Morrison 2003). Hence, union officials had to cooperate constructively with other agents. A trade union secretary of ver.di, the largest service sector union in Germany, at the local administration in Kassel stresses:

> To gain access to the temps in a firm, it is indispensable to act jointly with a dedicated, strong works council. In companies where works councils ignore the concerns of the temps and prefer to avoid further conflicts with the management and prefer to concentrate on the issues related to the permanent staff or where no codetermination structures exist at all, you have no reasonable chance to assist the temps. (Translation of the citation: D. Meyer)

The architecture had to be receptive of new routines emanating from the bottom up, and had to integrate external routines from works councils at both temporary work agencies and at companies employing temps. Interestingly, what we found was that the place and way of learning at the sub-national level to a very low extent depend *on the unique labour market structures of the particular region*, e.g. on the number of temps as percentage of all employees. Rather, important impulses for change often stem from the actions of agents, here the union officials. A union secretary, who was responsible for the regional campaign for temps in the IG Metall district North Rhine-Westphalia, highlights the importance of active secretaries:

> Whether it is in the Bavarian Forest, in the Rhine-Main area or in the Eastern federal states – no temp enters a trade union unless there is an ambitious union secretary who addresses his concerns actively and presents solutions to his or her problems. (Translation of the citation: D. Meyer)

There is normally a close relationship between union secretaries and works councils, at both a formal and informal level. Furthermore, unionists and members of works councils must be in close contact with the temps. Thus, we see that a combination of the person-oriented and organizational approaches is analytically useful.

The loss of members and the large percentage of temps in the new federal states strongly suggest that unions should make a priority to organize temps in East Germany. In Leipzig which is a telling example,there were 9,520 temporary workers, mainly in the automotive industry who made up 4.8 per cent of the work force in June 2007. The local administration office is very active, and one of the most impressive successes of the learning processes of the IG Metall Leipzig is the signing of tariff agreements with companies employing temps. The pay agreements they were able to reach guarantee temps a higher wage than they would have been entitled to under the conditions of the region-wide tariff agreements for temps. This way the discrepancy between their income and that of those permanently employed in firms employing temps is reduced at the local and regional level.

However, as argued above, the socio-economic environment, i.e. the large proportion of temps, is not decisive of trade union strategies aimed at organizing temps. At the local administrative offices in Erfurt, for instance, where temps numbered 6,300 and made up 6.6 per cent of the work force in June 2007, the local administration did not recognize a need to take action (Federal Employment Office 2008).

There are no significant differences in union strategies towards temps along the East-West divide and the North-South divide in Germany. In the same vein, whether the local administrative office of the trade union is responsible for an urban or more rural area does not necessarily affect the extent to which temps are recruited. A high density of temps in the Ruhr Area in North Rhine-Westphalia may result from the large percentage of jobless people in this 'old industrialized' area. Furthermore, a high density of temporary employment agencies and large numbers of unemployed are to be found in some locations, such as Wolfsburg, Emden and Oldenburg, where the automotive industry and its suppliers are located and create a considerable demand for temps. Here we can also find trade union activities related to temps. Thus, it seems obvious that organizational agency plays a role especially in automotive regions, due to the specific progressive trade union culture which has developed in this part of manufacturing (Federal Employment Office 2008).

The Development of Dynamic Capabilities: IG Metall

Below we want to identify some of the bottom-up and top-down processes which may explain the creation of dynamic capabilities. The IG Metall is one of the largest unions in Germany. As mentioned above, the acceptance of temporary employment on the part of the DGB unions manifested itself in 2003 in tariff agreements. After the conclusion of these tariff agreements, a bottom-up process of dispersing the new routines began.

In some local administration offices foundations were laid for the unionization of temps at the national level. In the case of Berlin this started in 2004 where particularly ambitious persons in the local administrative offices functioned as promoters for the new routines (Dodgson 1993, Heery et al. 2004). The union secretaries in Berlin recognized the need for action and took the matter of organizing temps in their own hands. Among other things, they set up a working group and offered support for establishing works councils in temporary employment agencies.

The new routines were disseminated at the local level of the organization. For example, the head of IG Metall Bautzen, Stephan Henning, was given advice by his colleagues operating at the local level in Berlin, as he describes in the following:

> I asked them: 'How did you get the actions started? How did you get the works councils involved? How did you get the issue of malpractices in temporary work into the local media?' They had previously made up their minds about these questions and we took advantage of their experiences. Mutual exchange of experiences works well in the IG Metall. Not every local administration has to reinvent the wheel. (Translation of the citation: D. Meyer)

Hence, the development of dynamic capabilities at the local level was furthered by the close cooperation of union secretaries in different local administration offices.

Following the example of the local administrative offices, some district administrations adopted the new routines. Because of the higher level of authority and possibilities for exerting influence at the district level, broader tasks of representing the interests of temps were developed and local problem-oriented routines were modified. For example, regional initiatives are aimed at attaining the same wages for temps as for permanent employees.

The new approaches which were developed by some regional districts, e.g. Berlin-Brandenburg-Saxony, have spread to other districts. For instance, a trade union secretary who had launched a project for temps in the regional administration of Frankfurt in Summer 2007 deliberated with secretaries in the regional administration office of Berlin-Brandenburg-Saxony on how to implement their own campaign in the region of Frankfurt. Consequently, the new routines were disseminated at the district level.

After it became evident that the new strategies represent lasting dynamic capabilities, the executive board at the national level adopted the routines. The district projects were launched as a nation-wide membership drive directed at temps at a Union Convention of the IG Metall in 2007. The aims of the campaigns are the realization of the principles of equal pay and equal treatment, as well as achieving a high degree of organization among temporarily employed that would enable effective campaigns.

With this campaign, an attempt was made to disseminate the dynamic capabilities in the unions at various levels (multi-scalar) from the top down, and therefore an effort was made to create a superior architecture containing such new dynamic capabilities. According to a union secretary who co-manages the campaign at the federal board, the entire organization was requested to recruit new members from the ranks of temps:

> The goal of our campaign is to reach the point when temps are addressed in each and every single local and regional IG Metall administration in Germany. From north to south, from east to west. We pool all activities and thus gain momentum. (Translation of the citation: D. Meyer)

This final step in the learning process was intended to disseminate the dynamic capabilities throughout the union at all levels (multi-locationally). This case also illustrates the relevance of the organizational approach in explaining development of strategies to recruit temps.

Concluding Remarks and Perspectives

This study has shown that large organizations with a decentralized structure with many branches, such as trade unions, can deal with changes in socio-economic conditions with the dual approach of a general policy and locally and hierarchically differentiated strategies. The efforts at recruiting temps are a good example. We demonstrated that the multi-locational and multi-scalar developments first led to various dynamic capabilities within an organization, and then resulted in a new architecture of dynamic capabilities. In the end, a construction plan, i.e. a new strategy, was extended. This, the main actors expected would prepare the union for facing changes to the socio-economic conditions taking place and thus further its organizational development.

In this chapter we have discussed the explanatory relevance of the structural approach, the organizational approach and the person-oriented approach and argued that it varies according to spatial scales and organizational levels. The structural approach can explain the general strategy of the executive board of the trade union. However, in order to understand multi-locational and multi-scalar dynamic capabilities, the person-oriented approach seems to be more adequate. However, individual agents can only succeed if they are embedded in beneficial, constructive relations with their colleagues and partners. Thus, the organizational approach, focusing mainly on network relations, highlights the social conditions and conducive personal relations for agents implementing new routines.

We have pointed out the effectiveness of such multiple learning processes. It should be emphasized that this is based on observations of recent learning processes up to the present. The many locations and the complex, wide-reaching structure of unions could possibly limit the diffusion of dynamic capabilities and turn an advantage into a disadvantage. Not every agent is a promoter, but could also turn out to be a gatekeeper who prevents the creation of innovative practices. Whether some district and administrative offices will be permanently cut off from learning processes cannot be determined at present. The development of dynamic capabilities at an organization-wide level also requires efficient top-down processes such as, for example, the power of persuasion.

Considering the reasons why dynamic capabilities have developed in certain places and not in others, we have emphasized three aspects: structural pressures, agents as promoters and agency in network relations. We saw that structural influences are important for the overall trade union strategy. At the same time, the promoters and the organizational context in which they are embedded play a role at the decentralized level. Although union organizations are more strongly bottom-up oriented than companies, such a view could also be of interest to the on-going debate on decentralized learning processes in subsidiaries of foreign direct investments. Thus, a focus on the development of dynamic capabilities highlights how unions may tailor their responses to changing socio-economic environments.

Our findings derive solely from developments in Germany. However, temp agencies are an increasingly significant employer in many countries, and the loss of members is affecting unions in many countries. Thus, a number of trade unions in other countries than Germany have also accomplished similar learning processes and implemented projects for the organization of temps. Some Dutch unions have established organizational subunits that are responsible for temps, for example the FNV Flex which is part of the largest private sector union FNV Bondgenoten. The Belgian General Federation of Labour has implemented an annual 'interim day' to attract public interest to the working conditions and rights of temps (European Foundation for the Improvement of Living and Working Conditions 2008).

When comparing the strategies of trade unions in other countries with those examined in Germany, one ought to bear in mind that the institutional setting in each country differs significantly. Trade unions are embedded in specific national contexts and arrangements of social dialogue and collective bargaining at the national level differ widely. Thus the observed developments in Germany are not necessarily transferable to other countries.

References

Amin, A. and Cohendet, P. 2004. *Architectures of Knowledge*. Oxford: University Press.

Argyris, C. and Schön, D. 1978. *Organizational Learning*. Reading: Addison Wesley.

Behrens, M., Fichter, M. and Frege, C.M. 2003. Unions in Germany: Regaining the initiative?, *European Journal of Industrial Relations* 9, 25–42.

Callon, M., Millo, Y. and Muniesa, F. (eds), 2007. *Market Devices*. London: Blackwell.

Castree, N., Coe, N., Ward, K. and Samers, M. 2004. *Spaces of Work.* London, Thousand Oaks, and New Delhi: Sage Publications.

DGB (Deutscher Gewerkschaftsbund) 2009. *Mitgliederzahlen*. Available at: www.dgb.de/dgb/mitgliederzahlen/gesamt1950_1993.htm/. [Accessed 18.03.08].

Dodgson, M. 1993. Organizational learning, *Organization Studies* 14, 375–94.

Dølvik, J.E. and Waddington, J. 2004. Organizing marketized services, *Economic and Industrial Democracy* 25, 9–40.

Dörre, K., Kraemer, K. and Speidel, F. 2006. The increasing precariousness of the employment society. *International Journal of Action Research* 2, 98–128.

Drinkuth, A., Riegler, C.H. and Wolff, R. 2003. Labour unions as learning organizations and learning faciliators, in *Handbook of Organizational Learning and Knowledge*, edited by M. Easterby-Smith and M.A. Lyles. Oxford: Oxford University Press, 446–61.

Druker, J. and Stanworth, C. 2004. The role of labour market intermediaries in workplace representation, in *The Future of Worker Representation*, edited by G. Healy, E. Heery, P. Taylor, and W.B. Heery. Hampshire and New York: Palgrave Macmillan, 229–44.

Eisenhardt, K.M. and Martin, J.A. 2000. Dynamic capabilities. *Strategic Management Journal* 21, 1105–21.

European Foundation for the Improvement of Living and Working Conditions 2008. *Temporary Agency Work and Collective Bargaining in the EU*. Available at: www.eurofound.europa.eu/docs/eiro/tn0807019s/tn0807019s.pdf [Accessed 06.03.09].

Federal Employment Office 2008. *Statistik Arbeitnehmerüberlassung*. Available at: www.pub.arbeitsamt.de/hst/services/statistik/200706/iiia6/aueg/aueg-zrd.xls [Accessed 18.03.08].

Federal Employment Office 2009. *Der Arbeitsmarkt in Deutschland*. Available at: www.pub.arbeitsamt.de/hst/services/statistik/200806/iiia6/sozbe/zeitarbeitd.pdf [Accessed 05.03.09].

Fuchs, M. 2008. Subsidiaries of multinational companies, *Geography Compass* 3, 1–12.

Heery, E., Healy, G. and Taylor, P. 2004. Representation at Work, in *The Future of Worker Representation*, by G. Healy, E. Heery, P. Taylor, and W.B. Heery. Hampshire and New York: Palgrave Macmillan, 1–36.

Herod, A. 1998. The spatiality of Labour Unionism, in *Organizing the Landscape. Geographical Perspectives on Labour Unionism*, edited by A. Herod. Minnesota: University of Minnesota, 1–36.

Herod, A. 2003. Workers, space and labor geography. *International Labor and Working-Class History* 64(Fall), 112–38. [Special issue on 'Workers, Suburbs and Labor Geography'.]

Jonas, A.E.G. 1996. Local labour control regimes. *Regional Studies*, 30, 323–38.

Lier, D. C. 2007. Places of work, scales of organising. *Geography Compass* 1, 814–33.

Martin, R. and Morrison, P. 2003. *Geographies of Labour Market Inequality*. London: Routledge.

Massey, D. 1984. *Spatial Divisions of Labour – Social Structures and the Geography of Production*. London: Macmillan.

Peck, J. 1996. *Work-place. The Social Regulation of Labour Markets*. New York: Guilford Press.

Phelan, C. 2007. Worldwide trends and prospects for trade union revitalisation, in *Trade Union Revitalisation. Trends and Prospects from 34 Countries*, edited by C. Phelan. Oxford: Peter Lang, 11–37.

Schreyögg, G. and Kliesch, M. 2003. *Rahmenbedingungen für die Entwicklung organisationaler Kompetenz*. Berlin: Quem Materialien.

Teece, D. and Pisano, G. 1994. The dynamic capabilities of firms, *Industrial and Corporate Change* 3, 537–56.

Vitols, K. 2003. Die Regulierung der Zeitarbeit in Deutschland, *Duisburger Beiträge zur Soziologischen Forschung* 5.

Chapter 9
Union Power and the Formal–Informal Divide

Gunilla Andrae and Björn Beckman

Introduction: Where do Good Institutions Come From?

What makes societies capable of delivering social welfare to their members? This chapter explores the potential for organizing across the formal–informal divide to enhance welfare and its wider implications for industrialization and development. The issues are approached from the formal end, reflecting primarily our own earlier studies of trade unions that suggest that these have a special role by virtue of their insertion in the political economy (Andrae and Beckman 1998, Beckman 2004). We begin by explaining this source of union power. Formal wage workers, however, are few in comparison with those in the informal economy, especially in Africa, and trade unions depend on their ability to engage in alliances. We ask what benefits accrue to those in the informal economy from such organizing across the formal–informal divide. Some international organizations have far reaching expectations of the benefits from extending the achievements of formalization, as in the case of the 'Decent Work' campaign of the ILO (2002) or in the UN Commission on the Legal Empowerment of the Poor (2008). In the latter case, an overall increase in the demand for goods and a greater efficiency of production are expected from the diffusion of the gains of the formal economy. This may well be. The focus in our case, however, is the politics of influencing the state and the direction of development policies.

What is the scope for organizing across the formal–informal divide? The Nigeria Labour Congress (NLC) has repeatedly mobilized its allies against government policies especially over local petrol prices, in recent years through LACSO, the 'Labour-Civil Society Coalition' (Beckman and Lukman 2009). Market women, drivers, and students have featured prominently in such efforts. In this chapter we look at recent local attempts by the Nigerian textile workers union (NUTGTWN) to negotiate a merger with the associations of local tailors. The context is an alarming drop in local industrial production caused by global trade liberalization and most specifically, the end of the Multi-fibre Arrangement (MFA) in 2005 that has opened the flood gates for cheap imports from major producers in Asia, China and India in particular. The discussions of a merger of labour power in Nigeria across the formal–informal divide were prompted by expectations of immediate benefits on both sides, including increased membership for unions that have suffered a sharp decline and, on the tailors' side, for enhanced status and better protection against harassment from tax collectors and market inspectors. Tailors, however, are small entrepreneurs with their own employees, apprentices and family workers and, as we have argued elsewhere (Andrae and Beckman 2007), there is no scope for the collective bargaining that is the hallmark of a union-based labour regime.

How can the achievements of trade unions be translated into benefits for society as a whole? What is the scope for protecting those achievements through organizing across the formal–informal divide? The failure of the Nigerian state to provide the basic preconditions for local production, including electrical power, fuel, water, and transport, has been exacerbated by the current (2008–09) downturn in the global economy. The collapse of economic infrastructure is a major concern on both sides of the divide. Some tailors believe that membership of the trade union will give them more political clout *vis-à-vis* the state, especially in delivering social services and social protection. On the trade union side there are also expectations that the tailors will benefit from being exposed to unions rights. The actual outcome needs to be explored empirically. However, in this chapter we are more concerned with the theoretical issues that guide our inquiry. Where do efficient institutions that are capable of delivering economic infrastructure, social services and protection come from? We argue that the organizations of civil society and their experience of regulating conflicts by constitutional means are a possible source of such 'institutional up-grading'. Organizing across the formal–informal divide, despite the apparent misfit in terms of the social relations of production, may contribute importantly in this direction.

In our earlier study of the textile and garment industry in Nigeria (Andrae and Beckman 1998) we noted that the union had successfully developed a *union based labour regime* as a means of regulating labour relations, extending it to new companies and environments that traditionally have been hostile to union influence. Organized wage-workers are few but strategically positioned and tend to pioneer workplace rights and collective bargaining. Our studies suggested that a union-based labour regime was diffused from a core of big multinationals and public enterprises to companies previously dominated by patriarchal and despotic labour regimes. The organizations of industry itself played a key role in this disciplining process, dominated as they were by 'respectable' companies that were anxious not to be undercut by those who were free to suppress wages as well as working conditions.

Constitutionalism, as we see it, is a way for society to regulate conflicts of interests. Workplaces are rife with conflicts over the ultimate distribution of income, rights of representation, strikes, disciplinary procedures, and working conditions in general. At one level it may look as a 'zero-sum-game' where a loss on one side is a gain on the other. However, as documented in the Nigerian case, employers also benefit from workers being organized and represented. It allows for predictable discussions and organized bargaining rather than the possibility of workplace anarchy and violence, along with their concomitant costs of repression and surveillance. Employers in the Nigerian textile industry, we observed, were willing to concede union rights in exchange for a more disciplined work force expecting, of course, that not only productivity but also profits would be enhanced. Constitutionalism in this sense is thus one way of managing conflict in the workplace based on agreed and negotiated rules where unions serve as an active party representing the workers. The outcome is primarily the result of successful contestation on the side of the workers of the despotic or unilateral regimes imposed by the employers. It is neither permanent nor irreversible but keeps being affected by the balance of social forces in society at any time, including the way in which the state intervenes, its laws, and its composition.

In other parts of the world, including parts of Africa, autonomous workers' organizations are repressed both by the state and the employers. However, there is no inevitable logic to such repression and we argue that capitalist development is capable of accommodating to very different types of labour regimes. This view draws support from the experiences of the Nordic 'Welfare States', where trade unions have provided core support for Social Democratic governments. In our case, however, the argument is developed from what has been observed in the Nigerian context. How can these undoubted achievements be protected? Can they serve as a basis for transforming institutions and the direction of the state? What role can organizing across the divide play in reinforcing this possibility?

Institution-building and the State

Central to our understanding is the critical role of organizations in civil society in institution-building from below, drawing on their competence in regulating conflicts in the areas where they are located. Constitutionalism, that is rule-regulated management of conflicts of contradictory interests, is closely linked to institutional capacity. The process is often spearheaded by trade unions in view of their experience of rights, including the right to organize, voice grievances, elect representatives and engage in collective bargaining. The specific insertion of formal wage-workers in development gives them self-interest in protecting and expanding wage employment, that is, in social development, industrialization, and modernization in general. It gives trade unions a particular stake in the competence of state institutions that are supposed to provide the necessary preconditions for commodity production and social policy. They develop a vested interest in the politics of institutional up-grading. Urbanization and the growth in formal wage-work affect the reproduction of labour power outside the work place. It is increasingly less possible to fall back on the reproductive logic of the peasant economy. Although some workers may retain claims to land in the village, rural communities have their own problems of self-reproduction and cannot necessarily accommodate redundant or retried wage-earners, especially if they have been cut off from rural production for long. Organized workers therefore also develop a vested interest in social policy, in the efficient and cheap provision of social services, including education, health, electricity, and water supply, as well as in 'social protection', that is, in collective solutions, either at the workplace or through state spending, to the problems of child-rearing, unemployment, and retirement.

The importance of such institution-building from below is particularly great in societies, as in much of Africa, where state institutions and ruling classes are weak and ineffective due to the limited and distorted history of state and class formation, including the exposure to colonialism and imperialist domination. The Nigerian state is penetrated by clientelist relations that tend to paralyse the growth of institutions capable of responding to the expressed development needs of its citizens. The state makes repeated attempts to establish control over such social forces, including the labour movement, that are not effectively integrated and subordinated by these clientelist networks. Unions in their turn struggle to ensure state backing. While often operating without authorization from laws that specify, for instance, when they are permitted to go on strike and not, unions still need the legal support of the state, not the least when asserting union rights against recalcitrant employers.

In Nigeria forms of labour organization existed in strategic wage-earning sectors as in harbours, railroads, and public works at an early point. It was only in the 1930s, with the threat of an approaching war, that the colonial government sought to regulate the labour scene, introducing a 'Labour Ordinance' and intervening actively in the work place (Hashim 1994). During the political turmoil of the early post-war period, trade unions played an active part in the politics of the anti-colonial movement and new laws were introduced by a labour-led colonial government. Current labour legislation is largely a product of the period after Independence (1960) and the Civil War (1968–70). A military government in the late 1970s imposed a unified trade union structure with one national centre (as distinct from several competing ones) and a structure of industrial unions, each with an organizational 'monopoly' within its specified domain. Despite repeated efforts to intervene in support of more government-friendly elements, the Nigerian state has failed to ensure its control. In 2005 the Obasanjo government revised the law to allow for more union centres. In practice, the Nigeria Labour Congress (NLC), as constituted in 1978, has been only marginally affected and continues to dominate the labour scene. The textile workers' union (NUTGTWN) discussed in this chapter has been central in giving political and financial support to the NLC. Since the turn of the century, however, the union has suffered from the sharp decline of the textile industry (Aremu 2006). This is a time when unions have taken a renewed interest in their relations to the informal economy.

Organizing in the Informal Economy[1]

At the level of the NLC it has long been realized that the ability of the unions to influence government policy depends on the mobilization of their allies in civil

1 Our own background in studying the informal economy goes back to our earlier work on Ghana's industrialization (Andrae 1981) and has been a theme in the 'People, Provisioning, and Place' programme in the Department of Human Geography at Stockholm University. In 2007, a major conference on 'Informalizing Economies and New Organising Strategies in Africa' was organized at the Nordic Africa Institute, Uppsala, by Ilda

society. In the past essentially middle-class and professional groups, like the National Association of Nigerian Students (NANS) and Women in Nigeria (WIN) have been close associates (Lukman 2005, Ibrahim and Salihu 2007). Simultaneously, a range of civil society groups and human rights activists have been anxious to seek out the NLC both in search of protection for themselves against a repressive state and in the pursuit of a radical agenda within the popular democratic movement. The collapse of the military dictatorship in 1998 and the re-establishment of party government in 1999 saw an outbreak of union-led civil society activity (Beckman and Lukman 2009). However, the organizations of market women, craftsmen, and traders, although critical in providing popular support for NLC campaigns against government attempts to 'deregulate' and raise local petrol prices, were conspicuously absent from these attempts of forming a 'Labour-Civil Society' network. It is only in recent years, largely through influences from the ILO and from trade unions elsewhere, for instance in Ghana and South Africa, that groups in the informal economy have been considered as potential partners in more formal alliances. Being preoccupied with their own organizational achievements, unions are often ignorant (and so are union-based scholars!) that these groups often have a long history of intricate organizing of their own, including executive officers anxious to protect their positions, the constitutional rules of their associations, and their own agenda. All of this, not just differences in social relations, will have an impact on the propensity of informal 'workers' and their organizations to see unions as a suitable partner for organizing across the formal–informal divide.

As in much of Africa, a large part of Nigeria's population finds its livelihood in small-scale activities that remain outside the reach of state regulation concerning conditions at work, in their social lives and regarding conditions of negotiation and organization. Efforts at extending the basic international labour standards and social security systems to these groups are high on the agenda of international organizations. They are pioneered by ILO with its Decent Work strategy from 2002, based on arguments that these conditions cannot be reserved for formal wage-earners alone, especially in view of the dominant role of the informal economy in the South (ILO 2002). Recently the UN Commission on Legal Empowerment of the Poor (UN 2008), has focused on the need to extend the legal rights of informal workers and entrepreneurs and thus on finding ways of formalising the informal economy largely from above. In this strategy legalization is intended to cover also rights of organization. Others (Chen 2005, Gallin 2001 and Horn 2007) in probing the scope for joining formal and informal systems of regulation, however, emphasize the structural limitations to such strategies. They have rather focused on ways that informal workers' organization may be promoted to voice their interests and fight for their own type of constitutionalism.

Lourenço-Lindell, a member of the group (Lourenço-Lindell 2007). We also contributed a paper (Andrae and Beckman 2007) to this conference, with a focus on the relations between unionized workers in the textile industry and workshop-based tailors in markets and commercial areas, to which the present one is a follow-up.

Such agendas of informal organizing are pioneered by the organization WIEGO (Women in Informal Employment Globalizing and Organizing) with support of other international organizations (e.g. the Global Labour Institute, GLI) and national and local NGOs.

There is, thus, wide agreement that organization is a crucial prerequisite for acquiring the voice to influence conditions of work and whatever social benefits that may also be negotiated through relations in the workplace. At the same time there is awareness that there are structural factors behind the weak bargaining power particularly of some of these groups, depending on the way they are inserted into political and economic contexts. The informal economy covers a wide variety of social relations. It contains, as Martha Chen notes in her paper 'Rethinking the Informal Economy' (2005), on the one hand categories of self-employed, with or without their own employees and including home-based, market- or street-based workers or those based in workshops in commercial areas. On the other hand, it contains the employees of such informal entrepreneurs and, increasingly, those working in formal establishments as casual workers. We may further distinguish between those who work for other producers and those who produce for final markets. In the former case we need, as Lund (2005) does, to situate the informal producers as part of a 'value chain' within which the counterparts for negotiation will have to be located. Informalization in that context is linked to the reorganization of the formal economy itself, involving subcontracting and outsourcing, as is the case of many small-scale tailoring workshops in South Africa.

In the case that we focus on in this chapter, the Kaduna tailors in Nigeria, these are typically found in workshops in markets or commercial areas and oriented towards final consumption. They are small entrepreneurs working with apprentices, family workers, and sometimes a few wage working employees (Andrae and Beckman 2007). As wage work in the textile and other industries is contracting, sectors like tailoring become major sites of expanding informalization in this context.

The scope for organizing differs between the different strata. Two categories in particular tend to organize themselves; first of all those who operate as self-employed market and street traders, who join forces in relation to market and local authorities which regulate access to and use of local space, as e.g. Lourenço-Lindell (2008) describes for Maputu. Here the agglomeration in space of workers with similar interests and the same counterpart for negotiation lend them negotiating power. But also members of established crafts are likely to have organizations that represent workshop owners in their dealings with local authorities. In both cases the organizations commonly administer mutual benefits and support like credit and assistance in the case of funerals and sickness. Those employed by these entrepreneurs are on the other hand typically too scattered and too dependent on their employers, on account of paternalist relations and unstable working conditions, to form their own organizations. Their powerlessness is reflected in classically poor conditions, far from what the ILO would consider as 'decent'.

Those who focus on the importance of organizing at the informal end expect trade unions to contribute resources and skills to this effort (Gallin 2001, Pillay 2008). Some unions have been found to offer support against harassment from the local state. In the case of Ghana's TUC and street trader organization StreetNet as well as Accra market traders' organizations, trade unions serve as 'umbrellas' in support of training and awareness-raising as shown by Owusu (2007) and Gallin (2004). Other African examples of union support to informal organizations, e.g. from Zambia, are given by War on Want et al. (2006). International trade union federations feel a similar obligation to encourage their members to assist in organizing the informal economy (ITGLWF 2000).

Tailors and Textile Workers in Kaduna[2]

As we visited Kaduna in February 2006 and 2007 the Nigerian textile workers' union (NUTGTWN) was actively engaged in merger talks with the association of workshop and market based tailors. The textile and garments industry was at this time in rapid decline, in particular due to competition from Asia. A number of the large factories in this 'Manchester' of West Africa, located on the northern Nigerian savannah, had recently closed down. It was followed later in 2007 by the closure of the national flagship, UNTL, affecting some 4,000 directly employed workers and in addition 'tens of thousands of those on the employment of service-providers, contractors and distributors' (NLC 2007). It became a national issue and the NLC and the smaller federation, TUC (mostly senior staff), joined in an appeal to the President of Nigeria to do something about the 'rapidly declining fortunes of the Nigerian textile industry'. The disaster had been aggravated by large-scale smuggling with 'connections at the highest level of the government'. The industry also suffered from 'infrastructural bottlenecks' that were making domestic production unviable and uncompetitive. In particular, the unions pointed to the shortages and scandalous costs of 'black oil', the fuel on which the industry depends. A 'Textile Sector Revival Fund', set up under President Olusegun Obasanjo, had made big promises, but failed to deliver (NLC 2007, 2009a). Only in early 2009, as the crisis of the textile industry was reinforced by the general downturn of the economy, did the Yara'dua government announce the 'quick and immediate release' of the funds (NLC 2009b).

At first, the effect of the crisis was clearly to bring unions and tailors closer together. From the point of the textile union, the immediate purpose of the attempted merger was to boost its membership at a time when the industrial collapse had sharply undercut its standing in society, not least in the politics of the NLC, the national body, where the textile union had played an influential role. While the move had its strong supporters among the tailors it also met with hesitation in some

2 *Sources:* The chapter draws on discussions with unions, textile workers, and tailors' associations in Kaduna and Kano, Nigeria, in February 2006 and 2007.

quarters. They had their own organization for negotiating with the authorities and they looked to the state for marketing support and small scale industry subsidies. The type of negotiations and collective bargaining that takes place within formal industry was seen as irrelevant to the tailors while many of the issues that concerned them the most were outside the agenda and competence of the textile union. Some tailors felt that they had developed their own methods for avoiding the harassment by local authority officials and tax collectors. For others membership of the trade union was seen as an important way of enhancing their authority and standing even if there were limits in terms of the immediate benefits that would accrue to them.

In a previous paper (Andrae and Beckman 2007) we suggested that a merger may not necessarily be a good idea in view of the wide differences in social organization and labour relations. Whatever forms of co-operation and alliances that were agreed had to recognize these differences and had to build on the experience of autonomous organizing on both sides. Such alliances across the formal–informal divide have their own rationale without necessarily assuming that those on the informal side would benefit from being exposed to the superior modes of organizing on the formal side. Even if such an impact should not be excluded, it is not, as one may get the impression from the ILO discourse, primarily a matter of unions taking a lead in 'extending formalization' to the informal economy. Rather unions also need wider popular support. In the case of the textile union this has taken on additional urgency with the decline of the textile industry which makes the workers more vulnerable. As demonstrated by protracted and endless court cases companies, when closing down factories or moving businesses overseas, feel free to ignore agreements with the union over retrenchments and compensation. Both states and companies take advantage of the industrial downturn in an attempt to role back a union-based labour regime. Unions need to reach out to the informal economy in order protect their achievements. So, what is in it for those on the informal side of the divide?

Both sides expected that co-operation would enhance their political clout. On the union side the immediate interest was in increasing its membership at a time of industrial downturn while tailors wanted trade unions to protect them against the harassment of corrupt market inspectors and tax collectors. The common interests, however, were more basic. Just as the unions, the tailors we spoke to were dismayed by the failure of the state to provide the basic infrastructure necessary for production. Both sides were negatively affected by the repeated, often daily, power failures, insufficient and irregular supply of water, and the poor state of public communications. Even if differently situated in the commodity chain, tailors and textile workers had a common interest in the orderly regulation of trade, especially in the need to impose restrictions on the importation of second-hand and ready-made clothing that undercut local production. Now even leading politicians and officials in the government felt free to flout the official ban and engage in massive smuggling. Tailors and unionized factory workers had a common interest in the institutional upgrading of society. The economic downturn, which in the Nigerian case was particularly felt in the global collapse of petrol prices in 2008–09, makes co-operation across the formal–informal divide even more urgent.

A frontier in the ILO vision of development is that social protection, such as welfare services, pensions and insurance, will be extended to the poor masses of the world. The 'Decent Work' policies include apart from conditions at work also conditions of social provision and protection (ILO 2002). Social policy has increasingly come into focus with particular attention to the politically weaker groups. In the spate of recent studies on social policy and protection among international organizations (ISSA, UNRISD, GLI, WIEGO) some have probed the limits to coverage among those based in the informal economy. The potentials of drawing on formal economy resources for extending coverage also to them have been a particular concern (Chen 2005) among interesting work on documenting experiences of independent organizing in the informal economy (see for instance Lund and Srinava 2000). Again, trade unions are assumed to have a special mission in view of their demonstrated capacity to include the reproductive issues in collective agreements, including medical attention, food- and housing allowances, and retirement benefits. It is not clear how the special mission in this respect will be affected by the economic crisis or, in the African context, by the industrial collapse. An even greater share in the costs of reproducing the labour force is likely to be carried by families and communities. Access to land and informal work opportunities will gain in importance. However, the demand for the cheap and orderly provision of services in education, health, and water is unlikely to abate and this is where unions seem to have a special mission in the defence of social service provision and social protection.

The provision of social services opens areas for organized co-operation, tapping commonalities of interest across the formal–informal divide. As the bargaining power of the industrial union is weakened, its ability to negotiate favourable deals with employers over social benefits, like health services, and allowances for meals, housing, and transport, is undermined. In the case of medical attention, for instance, the union had obtained favourable conditions in most textile factories not just for the workers but for their families as well. Allowances have contributed importantly to the value of the 'take-home' pay. In the current atmosphere of economic crisis and industrial collapse these achievements are unlikely to be adequately protected. Also the workers' capacity to pay for the increasingly privatized and costly supply of water, electricity and waste management is undermined.

As the factory-based income and benefits deteriorate, more importance is likely to be attributed to the terms under which services are provided outside the factory, private and public. In the attempt by the union to protect the interests of the workers, the social policies of states and local governments are important not only to the welfare of the workers and unemployed but also to the survival and credibility of the union itself. As the social protection provided through factory work is weakened the conditions are increasingly similar to those in the informal economy, where low and erratic incomes have even less scope for absorbing the current increase in costs. Not that the 'workers' in the informal economy are entirely lacking social protection. In Nigeria, many are part of mutual savings schemes and can make claims in conjunction with ill-health and death, as stipulated, for instance, in the constitution of the Kaduna tailors' association (see Andrae and Beckman 2007). However, these features are marginal.

By joining the union Nigeria's tailors can not hope to access the social benefits that workers have negotiated from their employers, as part of their pay, nor can their employees. They may however expect to enhance their ability to influence public policy on the supply of basic social services, like water, health and education. The cheap and orderly provision of such services is high also on their agenda. It gives the activists on both sides a strong enough case for organizing across the formal–informal divide.

Conclusions: Protecting 'Union Power'

Alliances of union and labour power across the formal–informal divide may serve different purposes. The chapter has discussed the importance of such alliances in enhancing union influence on the direction of the state. In a country like Nigeria, the state lacks strong foundations either in an historical bureaucracy or in the formation of contemporary ruling classes. Its institutions are penetrated and fragmented by clientelist relations and foreign domination. Unions claim that they have a political alternative that is better rooted in popular-democratic aspirations, including a vision of national development and modernization. The chapter goes into the basis of such claims, which it sees as a function of the manner in which formal wage-earners are inserted into the political economy of economically backward societies. The problem is that formal wage-earners, despite their aspirations, are few, especially in comparison with the vast numbers in the informal economy. Union influence on the direction of the state therefore depends on their ability to mobilize a wider support, including alliances across the formal–informal divide. Specifically, the chapter has looked at a merger attempt between the textile workers union and the local associations of tailors in Kaduna, a city in northern Nigeria. While there are wide differences in the social organization of production, the two parties have common interests in ensuring that the state delivers in terms of economic infrastructure and social services. The problem is that the state is too weak and incompetent to do so. It requires the building of relevant institutional capacity. A major function of alliances across the formal–informal divide is therefore to contribute to the development of such capacity by voicing popular demands and by disciplining institutions from below.

References

Andrae, G. 1981. *Industry in Ghana. Production Form and Spatial Structure*. Uppsala: Scandinavian Institute of African Studies.

Andrae, G. and Beckman, B. 1998. *Union Power in the Nigerian Textile Industry. Labour Regime and Adjustment*. Uppsala: Nordiska Afrikainstitutet; also

Somerset, NJ: Transactions Publ. and Kano: Centre for Research and Documentation, 1999.

Andrae, G. and Beckman, B. 2007. *Alliances Across the Formal–informal Divide: South African Debates and Nigerian Experiences*. Paper to conference on Informalizing Economies and New Organising Strategies in Africa. Uppsala: Nordic Africa Institute, April 2007. As revised.

Aremu, I. 2006. *Tears Not Enough*. Lagos: Frankad Publishers.

Beckman, B. 2004. Trade unions, institutional reform and democracy: Nigerian experiences with South African and Ugandan comparisons, in *Politicising Democracy: The New Local Politics of Democratisation*, edited by J. Harriss, K. Stokke, and O. Törnquist. Basingstoke: Palgrave.

Beckman, B. and Lukman, S. 2009. The failure of Nigeria's Labour Party, in *Trade Unions and Party Politics: Labour Movements in Africa*, edited by B. Beckman, S. Buhlungu, and L.M. Sachikonye. Forthcoming. Cape Town: HSRC Press.

Chen, M.A. 2005. *Rethinking the Informal Economy: Linkages with the Formal Economy and the Formal Regulatory Environment*. Research Paper No. 2005/10. Stockholm and Helsinki: EGDI and UNU-WIDER Conference on 'Unlocking Human Potential: Linking the Informal and Formal Sectors', Helsinki.

Gallin, D. 2001. Proposition on trade unions and informal employment in times of globalisation, in *Place, Space and New Labour Internationalisms*, edited by P. Waterman and J. Wills. Oxford: Blackwell.

Gallin, D. 2004. *Organizing in the Global Informal Economy*. Paper to Bogazici University Social Policy Forum: Changing Role of Unions in the Contemporary World of Labour, Istanbul.

Hashim, Y. 1994. *The State and Trade Unions in Africa: A Study of Macro-Corporatism*. The Hague: Institute of Social Studies.

Horn, P. 2007. *Informal Employment. Voice Regulation on the Informal Economy and New Forms of Work*. Available at: http://www.globallabour.info/en.

Ibrahim, J. and Salihu, A. (eds), 2007. *Feminism or Male Feminism? The Lives and Times of Women in Nigeria (WIN)*. Kano: CRD.

ILO 2002. *Decent Work and the Informal Economy*. Report VI to the International Labour Conference, Geneva.

ITGLWF 2000. *Global Solidarity in Global Industry: An Agenda for ITGLWF Action*. 8th World Congress of the International Textile, Garment and Leather Workers' Federation. Norrköping, 26–30 June 2000.

Lourenço-Lindell, I. 2007. *Introduction. Informalising Economies, New Challenges and Organised Responses*, Paper to the conference on Informalizing Economies and New Organising Strategies in Africa, Nordic Africa Institute, Uppsala, April 2007.

Lourenço-Lindell, I. 2008. Building alliances between formal and informal workers: experiences from Africa, in *Labour and the Challenges of Globalization: What Prospects* for *Transnational Solidarity?*, edited by A. Bieler, I. Lindberg, and D. Pillay. London: Pluto Press.

Lukman, S. 2005, Organisational dynamics and the NANC-NLC Alliance, in *Great Nigerian Students: Movement Politics and Radical Nationalism*, edited by B. Beckman and Y.Z. Ya'u. Kano and Stockholm: CRD and PODSU.

Lund F. 2005. *Informal Workers' Access to Social Security and Social Protection.* Background paper to UNRISD Beijing Plus 10 Report on Gender Policy. Sixth draft. Geneva: UNRISD.

Lund, F. and Srinava, S. 2000. *Learning from Experience. A Gendered Approach to Social Protection for Workers in the Informal Economy*. STEP and WIEGO. Geneva: ILO.

NLC 2007. *NLC and TUC Express Concerns over the Situation in the Textile Industry with Particular Reference to the Closure of UNTL, Kaduna.* Letter to the President of Nigeria, signed by the Presidents of NLC and TUC, 15 October 2007.

NLC 2009a. *New Year Message*, Abuja.

NLC 2009b. *N70 Billion Bailout for Textile Sector*. Press Statement, February 11.

Owusu, F. 2007. *The Strategies of the Ghana Trade Union Congress to Reach out to the Informal Economy*. Paper to the conference on Informalizing Economies and New Organising Strategies in Africa, Nordic Africa Institute, Uppsala, April 2007.

Pillay, D. 2008. Globalization and the informalization of labour: the case of South Africa, in *Labour and the Challenges of Globalization: What Prospects for Transnational Solidarity?*, edited by A. Bieler, I. Lindberg, and D. Pillay. London: Pluto Press.

UN 2008. *Making the Law Work for Everyone.* Vol. 1. Report of the Commission on Legal Empowerment of the Poor. New York.

War on Want, Workers Education Association of Zambia and Alliance for Zambia Informal Economy Associations. 2006. *Forces of Change: Informal Economy Organisations in Africa.* London and Kitwe: War on Want and Workers Association of Zambia, AZIEA.

PART III
Politics of Labour

Chapter 10
Between Revolutionary Rhetoric and Class Compromise: Trade Unions and the State

Herbert Jauch and Ann Cecilie Bergene

Introduction

The chapter explores the contextual embeddedness of trade unions through an analysis of the crisis experienced by the Namibian labour movement. A premise of this chapter is that labour, organized or not, should not be analytically separated from capital, the state and civil society, and, as a corollary, that trade union activity is necessarily political (Glassman 2004). Trade unions do not make independent decisions in the sense that their policies and strategies are formulated in a vacuum, rather they must be analysed in relation to the socio-political context.

This chapter aims at investigating theories on the involvement of trade unions in politics and hopes to further develop them through an examination of the involvement of trade unions in the Namibian struggle for independence. In order to achieve this, we will discuss theories on hegemony, the state, trade unionism and the ideology of social partnership. The empirical basis for this chapter is the relationship between a national union federation, the National Union of Namibian Workers (NUNW), and the state, dominated by the ruling party, the South West African People's Organization (SWAPO). We will examine the labour federations' role in promoting a working-class approach to politics. It is argued that while Namibia's trade unions still occasionally engage in radical rhetoric, they have accepted global capitalism as a given framework in which to operate without challenging its ideological and material base nor SWAPO's political commitment to its operation.

The NUNW played a prominent role during the liberation struggle and in the public policy debates after independence. Its history is in many ways similar to that of the Congress of South African Trade Unions (COSATU), as both were key agencies in terms of mass mobilization against apartheid and colonial rule (for a study of the former see Lier 2005). Like its sister unions in South Africa, the NUNW unions linked the struggle at the workplace with the broader struggle for political independence and formed links with other social and political organizations such as the churches, women's and students' organizations. Hence, the NUNW understood its role as that of a social movement, which could not address workers issues separately from those affecting the broader community. In the 1980s, the NUNW adopted a strategy of social movement unionism, as defined by Waterman (2001) and Munck (2002).

However, despite the role played by Namibian trade unions in the liberation struggle, and regardless of the fact that the labour movement is still among the strongest of Namibia's civil society organizations, trade unions seem to have lost much of their popularity and political influence in recent years. The development agenda of the Namibian government has become increasingly neoliberal, and the high levels of income inequality were perpetuated during the 20 years of independence (Central Bureau of Statistics 2008). The labour movement has only offered occasional resistance to neo-liberalism and has been unable to alter Namibia's socio-economic structures. We will argue that this is partly attributable to the role played by the trade union movement in the national-popular struggle for independence. Trade unions were ill prepared for the post-independence ideological turnaround of their former ally and now ruling party, SWAPO, and the notion of social partnership replacing working-class politics.

Namibia has about 30 trade unions split into two federations and several unaffiliated unions. The largest trade union federation is the National Union of Namibian Workers (NUNW), which represents 60,000–70,000 workers. The NUNW played a key role during Namibia's liberation struggle and continues to be affiliated to SWAPO. The second trade union federation is the Trade Union Congress of Namibia (TUCNA), which was formed in 2002 by unions that rejected the NUNW's party-political link (Jauch 2004). The TUCNA has 14 affiliates with a combined membership of about 45,000. The TUCNA unions focus predominantly on workplace issues and claim to be non-political.

After providing a brief history of the link between NUNW and SWAPO, this chapter sets out to develop a theoretical framework capable of explaining the historic events and current affairs in Namibia. In order to achieve this we will zoom in and out of the empirical events in Namibia, theorizing them as we move along.

The Historic Link between NUNW and SWAPO

The NUNW's history is closely linked to that of SWAPO as a result of the particular history of Namibia's liberation struggle. Namibian contract workers formed a central component of SWAPO in the party's formative years. The plight of contract workers was first taken up by the Ovamboland People's Congress (OPC) that was founded in Cape Town in 1957. Migrant workers in the Namibian compounds responded enthusiastically to the OPC, which expressed their aspirations. In 1958 the OPC became the Ovamboland People's Organisation (OPO) demanding 'political, social and economic emancipation of the people'. In 1960 the OPO was transformed into a national liberation movement – SWAPO. Its aim was to establish a unified, independent and democratic Namibia, free from colonial exploitation and oppression (see Moleah 1983, Katjavivi 1988, Peltola 1995). Following SWAPO's consultative congress in Tanzania, in 1969–70, several new departments were established within the party, including a Labour Department. Although the congress documents did not mention the formation of trade unions, a decision to establish the NUNW in exile was taken on 24 April 1970 (Peltola 1995). In 1978 the SWAPO Central Executive Committee decided to affiliate the NUNW to the World Federation of Trade Unions (WFTU), which provided a link between the NUNW and the socialist countries. In 1979 the NUNW set up its headquarters in Angola, under the leadership of John Ya Otto who served as SWAPO secretary of labour and NUNW secretary-general at the same time. Ya Otto prepared a constitution for the NUNW for adoption by SWAPO's National Executive Committee (NEC), but it was never approved. Some party leaders even responded negatively to the union initiative, fearing a strong and independent labour movement after independence (Peltola 1995). These early tensions between a potential working-class orientation of SWAPO versus a nationalist ideology were already decided in the run-up to Namibia's independence in favour of the latter. The NUNW unions were formally established from 1986 onwards and provided workers with an organizational vehicle through which they could take up workplace grievances as well as broader political issues, which were always seen as linked to the economic struggle. This occurred firmly within the SWAPO fold as the NUNW unions openly declared their allegiance to the liberation struggle and to SWAPO as the leading organization in the fight for independence. The NUNW unions enjoyed huge support even beyond their membership and played a critical role in ensuring SWAPO's victory in the elections of 1989 (Jauch 2007).

National-popular Struggles, Historical Blocs and the Building of Hegemony

In theorizing the involvement of the labour movement in national-popular struggles, the work of Gramsci is a valuable starting point. According to him, capitalist societies are a complex network of relations between classes and social forces which give rise to a plethora of organizations and institutions such as churches, political parties, trade unions, the mass media and a range of cultural and voluntary associations (Simon 1991). Of these institutions, the state apparatus stands out due to its monopoly of coercion. The capture of state power in what Gramsci (1930–32) terms a war of manoeuvre[1] has, according to Holloway (2002), been the dominant paradigm within most revolutionary movements. The idea is that once captured, the new revolutionary state can be used to change society, and in Gramscian terminology, the transition to socialism will require a war of position involving the expansion of the hegemony of the working-class throughout civil society (Simon 1991). Hence, Gramsci defined the war of position as complementary to the war of manoeuvre. While the latter is likened to 'frontal war' and involves the seizure of state power, the war of position is termed 'trench warfare' and pertains more to subjective dimensions such as ideology and hegemony (Sassoon 1987). According to Gramsci (cited in Simon 1991), in such struggles, new historical blocs are built. The concept of historical bloc is one of the more elusive ones in

1 This term has also been translated into 'war of movement'.

Gramsci's writings, and he uses it in two ways that have a common core (Sassoon 1987). We will only dwell on the second, since the first use is very abstract and concerns the relation between base and superstructure. In the second usage, historical bloc designates the relations between various (fractions of) classes, and is used to describe how different social forces relate to each other (Sassoon 1987). However, as pointed out by Rupert (1995), the concept of historical bloc should not be used interchangeably with that of political alliances of classes through which each class pursues its own interests. In Glassman's (2004) view, historical blocs may be regarded as coalitions between classes with (potentially) conflicting interests at one level. However, he emphasizes that they are more than that since they involve a temporary resolution of overt antagonisms into an acceptance of leadership based on consent. The basis for building such a bloc could be a common cause which overrides other (potentially conflicting) interests that the classes might have. Another important Gramscian concept is that of hegemony. According to Simon (1991: 23), hegemony 'is a relation between classes and other social forces' and a hegemonic class 'is one which gains the consent of other classes and social forces through creating and maintaining a system of alliances by means of political and ideological struggle'. Becoming a hegemonic class thus necessitates the building up of a system of alliances and taking account of the interests of other classes and social forces. Hence, when the hegemony of a class is established, the development and expansion of that class is 'conceived of, and presented, as being the motor force of a universal expansion, of a development of all the "national" energies' (Gramsci 1932–34a: 205). Gramsci (cited in Simon 1991) introduced the concept of national-popular to argue that a class cannot become hegemonic unless it transcends its own class interests and takes account of other popular and democratic struggles. The struggle for national independence is one of the examples listed by Gramsci. In Simon's (1991: 25) words, hegemony 'requires the unification of a variety of different social forces into a broad alliance expressing a national-popular collective will'. However, as we shall see, when the 'national-popular collective will' is bent on the short-term objective of attaining national independence and conquering state power, hegemony needs to be built all over again once it is achieved. This is why we will argue that Gramsci's theories need to be supplemented by a theorization of the state.

Fighting for the Independence of Namibia

The national liberation movement in Namibia is a good example in which the employment of a national discourse mustered the support of broad sections of the population, including peasants, workers, and parts of the petty bourgeoisie, to fight against foreign occupation and to seize state power. In such cases, a social force must retain its autonomy in order to defeat the enemy, but also create its hegemony in order to exercise leadership over other social forces (Sassoon 1987). Building hegemony in the Namibian civil society around the need to oppose the

South African regime, proved an easy task given the 'universality' of this interest and the basis of South African rule being coercion and not consent. The absence of consent in civil society meant that the concept of war of position did not become relevant until after SWAPO came to power and needed to build hegemony in order to secure its leadership. Hence, we agree with Ackers and Payne (1998) that the war of position is more relevant in capitalist societies that are ruled more by consent than coercion.

In the case of the Namibian liberation movement, SWAPO was born out of the struggle of Namibian workers against a highly exploitative migrant labour system and apartheid- colonialism. SWAPO's social base consisted predominantly of peasant farmers in northern Namibia and an emerging working-class in various towns. The movement was united around the broad goal of achieving national independence and abolishing colonial rule and labour exploitation. Hence, for Namibian workers inside the country the class struggle was intertwined with the struggle against racial discrimination and minority domination. The class struggle waged by workers was essentially seen as one and the same as the liberation struggle waged by SWAPO (Peltola 1995). Thus class differences were blurred and trade unions, membership and leadership alike, regarded themselves less as representing a particular class than as an integral part of a broader national liberation movement opposed to apartheid-colonialism. However, before independence, SWAPO claimed to play a vanguard role in the struggle 'of the oppressed and exploited people of Namibia. In fulfilling its vanguard role, SWAPO organises, unites, inspires, orientates and leads the broad masses of the working Namibian people in the struggle for national and social liberation' (SWAPO constitution of 1976, quoted in SWAPO 1981: 257). Likewise, SWAPO's political programme of 1976 was characterized by socialist rhetoric, inspired by the newly won independence of Mozambique and Angola, and by the support rendered by the Soviet Union to Namibia's liberation struggle. It stated that one of SWAPO's key tasks was: 'To unite all Namibian people, particularly the working-class, the peasantry and progressive intellectuals, into a vanguard party capable of safeguarding national independence and of building a classless, non-exploitative society based on the ideals and principles of scientific socialism' (SWAPO 1981: 275).

However, when SWAPO gained state power, it needed to build consent and hegemony throughout Namibian society around the way forward. This occurred at the expense of the revolutionary goals that had been set in earlier policy documents. Hence, once SWAPO became Namibia's ruling party in 1990, its acceptance of a capitalist order was rapidly consolidated. Revolutionary working-class politics were simply dropped while the capitalist structure of the economy was maintained. According to Magnusson (2008), SWAPO found itself in a situation in which it had to accommodate the diverging agential interests in the new Namibian state, among them the interests of the economic elite, consisting mainly of South African and international companies. Being still highly dependent economically on South Africa, SWAPO decided to leave foreign capital untouched in an effort not to disrupt the Namibian economy (Magnusson 2008). Additionally, the small Namibian business elite, which had been allowed to form under South African rule, also allied itself with SWAPO and was thus able to sustain itself. Hence, despite SWAPO having conquered state power and becoming the political elite, economic power remained mainly with foreign capital. Magnusson (2008) argues that SWAPO had to enter into alliances with this economic elite through which concessions had to be made in order to retain state power and remain the hegemonic class. These new alliances may have extended SWAPO's political influence into the economic domain, and, as argued by Simon (1991), such a combination of the leadership of a block of social forces with that of the sphere of production should be termed a historical bloc.

The NUNW maintained its links with SWAPO after independence through its continued affiliation. This link has led to heated debates both within and outside the federation. While the majority of NUNW affiliates argued that a continued affiliation would help the federation to influence policies, a choice we will return to below, critics have pointed out that the affiliation will undermine the independence of the labour movement and that it will wipe out prospects for trade union unity in Namibia. Trade unions outside the NUNW have charged that NUNW cannot act independently and play the role of a watchdog over government as long as it is linked to the ruling party. There is also a growing public perception that NUNW is merely a workers' wing of the ruling party, although the NUNW and its affiliates have on several occasions been vocal critics of government policies (Jauch 2007).

Trotsky (1969) argued that politically independent unions do not exist anywhere, and that trade unions should only strive for autonomy from bourgeois and petty-bourgeois parties but not for autonomy from the Communist Party. Trotsky (1969: 15) believed that 'the question of the relationship between the party, which represents the proletariat as it should be, and the trade unions, which represent the proletariat as it is, is the most fundamental question of revolutionary Marxism'. He argued that only the Communist Party could help the trade union movement to 'find its orientation' and that the Communist Party must win 'through the trade unions, an influence over the majority of the working-class' (Trotsky 1969: 21–22). At face value, it could thus be argued that the NUNW was correct in recognizing SWAPO as the revolutionary vanguard of Namibia's liberation struggle. However, it became clear that SWAPO regarded national independence, and not the proletarian revolution, as the primary goal of its struggle, and its achievement in 1990 required a redefinition of the role of trade unions. The function of political mobilization, which had taken centre stage in the years before independence, was taken over by SWAPO. Given the close structural links between the NUNW unions and SWAPO, as well as the fact that most union leaders played a prominent role in the party as well, there was a widespread expectation among workers that the SWAPO government would be a 'workers' government'.

However, due to the abovementioned emergence of a historical bloc containing such diverse agents as a foreign economic elite and poor Namibian peasants and workers, SWAPO was caught in a difficult limbo. Remembering the revolutionary promises during the struggle for independence and fearing that their former allies might challenge SWAPO's newly found power, the party started undertaking what Gramsci terms a passive revolution in order to forestall any consent spreading in civil society around the need for a socialist revolution.

The Capitalist State and Passive Revolution

In line with Poulantzas (1976) and Jessop (2008), we would argue that the state should be regarded as a social relation reflecting the balance of social forces, and that it could be analysed as a site, generator and/or product of strategies. This definition of the state brings with it certain consequences. According to Holloway (2002), social relations know no boundaries, but the tendency to assume that society means a national, state-bound entity lends a hand to the view that the state is a focal point in social transformation. In Holloway's view, social relations have never coincided with national boundaries and this necessitates abandoning the faith in conquering state power as part of a socialist revolution. Neither has history given us any evidence that this can successfully be done, nor, Holloway argues, is it a theoretically sound argument. While the state is legitimized through notions of sovereignty and being in a position to carry out the will of the citizens by exerting power in its territory, Holloway maintains that this isolates the state from its social context and grants it an autonomy that it does not enjoy. In his words, in reality:

> what the state does is limited and shaped by the fact that it exists as just one node -in a web of social relations … The fact that work is organised on a capitalist basis means that what the state does and can do is limited and shaped by the need to maintain the system of capitalist organisation of which it is part. Concretely, this means that any government that takes significant action directed against the interests of capital will find that an economic crisis will result and that capital will flee from the state territory. (Holloway 2002: unnumbered)

Holloway explains the above sentiment in the revolutionary movement with an instrumentalist perspective of the state. If the state is an instrument, or tool, to be wielded by the capitalist class during capitalism, this instrument, or tool, could, like a hammer, just change hands and be wielded by the working-class and then be harnessed in its favour. This, according to Holloway, fetishizes the state through abstracting it from the power relations in which it is embedded. Once this embeddedness is forgotten, the state is viewed as serving whoever wields state power. If we move beyond viewing the state merely as an instrument of a class, it is easier to get a glimpse of its class character, that is, its function in assuring the conditions under which a given class can develop fully albeit acting in the name of universal interests (Sassoon 1987). Governments today are thus trying hard to patch over any conflicting interests, and the task at hand, such as socio-economic development, is cloaked in terms of universal interests. Past discourses referring to delimited constituencies are binned in favour of an all-embracing national discourse, also rendering politics a much more 'neutral' matter.

As discussed above, hegemony presupposes 'that account be taken of the interests and the tendencies of the groups over which hegemony is to be exercised, and that a certain compromise equilibrium should be formed – in other words, that the leading group should make sacrifices of an economic-corporate kind' (Gramsci 1932–34b: 211). In other words, the political basis for hegemony must be rooted in class compromises and reforms whose intention is to concede material advantages to broad sections of the population (Sassoon 1987). The ability to lead hinges thus on both economic and ideological/political power. Although many would point to 'false consciousness' when explaining why the elite is able to muster consensus around this ideological hegemony, Jessop (2008) maintains that the support stems from the necessity of including popular-democratic demands, and certain (material) interests and aspirations of the 'people' in the hegemonic project. In order to achieve hegemony based on widespread consensus, the ruling class needs to be able to relate the activities of the state to the abovementioned 'illusory community' and its alleged common interests. The idea of a nation state rests on the construction of such a community with certain common interests, here national interests, which cut across class membership. For a consideration of the rise of modern nation-states and their solidity see Anderson's (1997) account of nations as imagined political communities arising out of the development of print-languages. The capitalist state has important class effects, and Jessop (2008) is particularly interested in how the state secures both the extra-economic and economic conditions for continued capital accumulation. What makes the state appear as a more or less unitary subject is, according to Jessop (1990), its role in promoting the common interests of an 'illusory community' through building hegemony. For instance, in its efforts to secure continued capital accumulation, the Namibian state has adopted neoliberal policies involving the establishment of Export Processing Zones (EPZs) and the introduction of privatization programmes. Opposition to such policies by the labour movement has frequently been countered by accusations from both capitalists and the government that trade unions are still living in the (ideological) past and that they are obstacles to economic growth and job creation (Jauch 2007). Capitalists and government alike have discursively coerced unions into accepting the concept of 'national competitiveness' through casting them as scapegoats for retarding 'national development' as defined by the hegemonic class (Bergene 2007a). Such arguments have even found resonance among some union leaders who are reluctant to support militant workers' actions and are torn between loyalty to the SWAPO government and their own members. Hence, Namibia today provides an example of what Gramsci termed 'bourgeois hegemony' where business interests are portrayed as constituting the 'national interest', accepted by subordinate classes, including significant sections of the trade union movement. Despite the desperate material situation of the majority of Namibia's working people (see for example Karuuombe 2002, Karamata 2006, Mwilima 2006, Shindondola-Mote 2008, LaRRI 2007), trade unions have failed to build a counter-hegemony. Instead, trade unions have been confined to a narrow 'economistic' struggle around 'bread and butter' issues, mostly in the form of collective bargaining.

However, hegemony is not reached once and for all, and is under continuous threat. Efforts to restore it in periods where it is questioned or resisted, through increasing or changing concessions granted to the 'people', is termed passive revolution by Gramsci (cited in Sassoon 1987). 'Passive' refers to the aim of preventing the adversary from developing through decapitating its revolutionary potential by, *inter alia*, absorption into traditional political organizations and reformism (Sassoon 1987). A key ingredient in the passive revolution is thus accepting certain demands from the broad masses while at the same time keeping economics and politics separate and restricting the working-class struggle to the economic terrain. This confinement is part of an attempt to preserve hegemony from being challenged, and the concessions given are accommodated within the prevailing social formation. By fostering technocracy through shifting the focus of discussions from principles to details, such as 'bread-and-butter' issues, and by supporting composite resolutions with a wording that embraces any conflicting interests and seeks a unity of opposites, the adversary is weakened through cooptation of the leading elements (Hyman 2005).

With the attainment of independence, the leading civil society organizations in Namibia were demobilized, and decision-making power shifted decisively towards party structures. As the leaders of the liberation movement became part of the state apparatus, they aligned themselves with the interests of both local and international capital. The secondary role allocated to trade unions and working-class interests was reflected in the way tripartism and social partnership became the cornerstone of labour relations after independence. Trade unions were expected to define a new role within this framework and although the NUNW had previously called for more radical change, it accepted the new framework with little resistance. The ideology and praxis of social partnership constitutes a consolidation of the ruling class' hegemony through ensuring the consent of the broad masses of the population and is an important instrument in passive revolutions (Bergene 2007b). By being invited to join policy negotiations, trade unions, and other civil society organizations, are made partly responsible for the outcome, which they then seek to justify, thus spreading consent in their ranks. Namibia's version of social partnership is essentially a reward by SWAPO for its working-class base that had played a decisive role in ensuring the election victory of 1989. However, social partnership does not represent a move towards granting labour a special status in the post-independence dispensation. Similarly, in the class compromises during the Fordist era in the Western capitalist countries, anti-capitalist parts of the labour movement were marginalized and repressed, and 'the social pact' resulted in depoliticization, deradicalization and bureaucratization of the union movement, and the ensuring improvements in wages and working conditions were 'what the labour movement gained in exchange for giving up its socialist project' (Wahl 2004: unnumbered). However, the notion of social partnership in Namibia is more of an ideological construct than a reflection of the social and economic balance of power, and serves to control and even weaken the labour movement. As argued by Kelly (1996), trade unions are not independent agents when it comes to goals, methods and resources since both the state and capitalists might constrain and suppress particular types of union behaviour.

Hence, although post-independence labour legislation in Namibia, including the Labour Act of 2007, has constituted a significant improvement for labour, it also served to reduce worker militancy by shifting the emphasis away from workplace struggles to negotiations between union leaders and management. Bargaining issues in Namibia were, and still are, narrowly defined and usually deal with conditions of employment only (Klerck et al. 1997). Furthermore, the vast majority of the working-class, the unemployed, informal economy workers, domestic workers, do not benefit from collective bargaining and thus still experience high levels of poverty. Moreover, according to Aglietta (1997), with globalization and the quest for global profitability, national collective bargaining ceases to be an important element in the system of national regulation, which is geared at meeting business demands for deregulation and competitiveness.

Trade unions, world-wide and in Namibia, have failed to mount a coherent challenge to the ideology of neo-liberalism, and the NUNW has tried to raise workers' concerns mostly through meetings with SWAPO leaders and government officials, and only on very few occasions resorted to more militant actions. The NUNW's strategy was thus based on lobbying while demobilizing its own membership to a large extent. As demonstrated above, unions are caught between what Dølvik (1997) calls the logic of influencing their external environment and decision-making processes and the logic of heeding the calls of the membership and retaining their legitimacy. As recognized by Bergene (2007a) this leads to a situation in which workers' initiatives, especially when militant, may be regarded as jeopardizing the position of union leaders as joint decision-makers, leading the latter in some cases to work against their own members in order to coerce them back into line. While union officials who partake in negotiations are made part of the hegemonic consensus in Namibia, the lack of performance in delivering real improvements for workers makes the grassroots much less easily co-opted. Not being geared in the same direction, initiatives by the rank and file are often frustrated by the national union leadership who do not want to support campaigns and actions that may threaten their own power base.

By Way of Conclusion

After 20 years of independence, Namibia's labour movement finds itself in deep crisis. Workers and their trade unions have had to realize that the changes after independence did not lead to the expected socio-economic transformation. Deep political divisions, not only between the NUNW and its rival federation the TUCNA, but also within the NUNW itself, worsen this dilemma. The NUNW displays deep-seated ideological contradictions. Sentiments of radical nationalism and liberation have survived and become mixed with an acceptance of neo-liberalism. As trade union leaders enter the corridors of power their views and interests increasingly converge with those of government and business. These developments point to a lack of clarity regarding the working-class base of the labour movement and whose interests it is meant to serve. Nationalist and 'populist' sentiments are dominant and trade unions hardly advance positions based on a class analysis. Namibian trade unions do not challenge capitalist social relations any more, and those unions who oppose party-political links do not base their position on a working-class ideology but merely claim allegiance to a 'non-political' trade union independence. This essentially amounts to accepting and increasing labour's narrow confinement to the economic sphere without challenging capital's hegemony through ideological and political struggles. What then are the options for the Namibian labour movement? Applied to Namibia, Gramsci's notion of working-class hegemony would have required the NUNW to engage in a new form of social movement unionism through which working-class interests could be articulated beyond the point of production in alliance with other socially excluded groups. However, the social movement unionism of the 1980s is no more than a historic memory and can hardly be repeated under the conditions of present-day Namibia. Also, neither unions nor NGOs in Namibia currently meet the essential requirements for successful social movement unionism such as a shared vision, internal democratic principles and openness for mutual exchange and co-operation. A return to class politics and transformative action would thus require internal changes within the labour movement as well as the conscious forging of new alliances. The involvement of the labour movement in the struggle for independence under the vanguard of SWAPO, and later in the new historical bloc, and SWAPO's later abandonment of working-class politics has put Namibian unions in a difficult position. While hegemony could easily be built in the Namibian civil society around the need for political independence from South Africa, the task of building a new hegemony around the way forward met with more difficulties as there were a plethora of agential interests and expectations to fulfil, coupled with the need for pragmatism. We have in this chapter argued that SWAPO's ideological turn-around and the developments in Namibia after independence can best be explained employing theories of the capitalist state and the Gramscian concept of passive revolution, and that trade unions must be analysed in relation to the socio-political context.

References

Ackers, P. and Payne, J. 1998. British trade unions and social partnership: rhetoric, reality and strategy. *The International Journal of Human Resource Management* 9(3), 529–50.

Aglietta, M. 1997. Postface to the new edition: Capitalism at the turn of the century: Regulation Theory and the challenge of social change, in *A Theory of*

Capitalist Regulation. The US Experience, edited by M. Aglietta. New Edition, 2000. London: Verso.

Anderson, B. 1997. The nation and the origins of national consciousness, in *The Ethnicity Reader: Nationalism, Multiculturalism and Migration.* Cambridge: Polity Press.

Bergene, A.C. 2007a. Trade unions walking the tightrope in defending workers' interests: wielding a weapon too strong? *Labor Studies Journal* 32(2), 142–66.

Bergene, A.C. 2007b. *Class politics, Hegemony and the Ideology of Social Partnership.* Paper presented at The Fourth Annual Historical Materialism Conference, 9–11 November, London.

Central Bureau of Statistics, National Planning Commission. 2008. *A Review of Poverty and Inequality in Namibia.* Windhoek: Republic of Namibia.

Dølvik, J.E. 1997. *Redrawing Boundaries of Solidarity? ETUC, Social Dialogue and the Europeanisation of Trade Unions in the 1990s.* ARENA/Fafo, Oslo.

Glassman, J. 2004. Transnational hegemony and US labor foreign policy: towards a Gramscian international labor geography. *Environment and Planning D: Society and Space* 22(4), 573–93.

Gramsci, A. 1930–32. War of position and war of manoeuvre or frontal war, in *A Gramsci Reader. Selected Writings 1916–1935*, edited by D. Forgacs. London: Lawrence and Wishart.

Holloway, J. 2002. *Change the World without Taking Power.* Available at http://libcom.org/library/change-world-without-taking-power-john-holloway. [Accessed 21.07.08].

Hyman, R. 2005. Trade unions and the politics of the European social model. *Economic and Industrial Democracy* 26(1), 9–40.

Jauch, H. 2004. *Trade Unions in Namibia: Defining a New Role?* Windhoek: FES and LaRRI.

Jauch, H. 2007. Between politics and the shopfloor: which way for Namibia's labour movement?, in *Transitions in Namibia. Which changes for whom?*, edited by H. Melber. Uppsala: Nordic Africa Institute.

Jessop, B. 1990. *State Theory: Putting the Capitalist State in its Place.* Cambridge: Polity Press.

Jessop, B. 2008. *State Power: A Strategic-Relational Approach.* Cambridge: Polity Press.

Karamata, C. 2006. *Farm Workers in Namibia: Living and Working Conditions.* Windhoek: LaRRI.

Karuuombe, B. 2002. *The Small and Micro Enterprise (SME) Sector in Namibia: Conditions of Employment and Income.* Windhoek: JCC and LaRRI.

Katjavivi, P.H. 1988. *A History of Resistance in Namibia.* Paris: UNESCO.

Kelly, J. 1996. Union militancy and social partnership, in *The New Workplace and Trade Unionism*, edited by P. Ackers et al. London: Routledge.

Klerck, G., Murray, A. and Sycholt, M. 1997. *Continuity and Change: Labour Relations in an Independent Namibia.* Windhoek: Gamsberg Macmillan.

Labour Resource and Research Institute. 2007. *The Namibian Wage Bargaining & Director's Remuneration Report 2006.* Windhoek: LaRRI.

Lier, D.C. 2005. *Maximum Working Class Unity? Challenges to Local Social Movement Unionism in Cape Town.* Master thesis submitted at the Department of Sociology and Human Geography at the University of Oslo.

Magnusson, O. 2008. *A Revolution Betrayed: How Namibian Class Relations are Played Out in the Work Regimes in the Retail Sector of Oshakati.* Master thesis submitted at the Department of Sociology and Human Geography, University of Oslo.

Moleah, A.T. 1983. *Namibia: The Struggle for Liberation.* Wilmington: Moleah publishers.

Munck, R. 2002. *Globalisation and Labour. The New 'Great Transformation'.* London: Zed Books.

Mwilima, N. 2006. *Namibia's Informal Economy: Possibilities for Trade Union Intervention.* Windhoek: LaRRI.

Peltola, P. 1995. *The Lost May Day: Namibian Workers Struggle for Independence.* Helsinki: The Finnish Anthropological Society in association with the Nordic Africa Institute.

Poulantzas, N. 1976. The capitalist state: a reply to Miliband and Laclau. *New Left Review* 95(January–February), 63–83.

Rupert, M. 1995. *Producing Hegemony: The Politics of Mass Production and American Global Power.* Cambridge: Cambridge University Press.

Sassoon, A.S. 1987. *Gramsci's Politics.* Second Edition. Minneapolis: University of Minnesota Press.

Shindondola-Mote, H. 2008. *The Plight of Namibia's Domestic Workers.* Windhoek: LaRRI.

Simon, R. 1991. *Gramsci's Political Thought: An Introduction.* Revised Edition. London: Lawrence & Wishart.

SWAPO 1981. *To Be Born a Nation. The Liberation Struggle for Namibia.* London: Zed Press.

Trotsky, L. 1969. *Leon Trotsky on Trade Unions.* New York: Pathfinder.

Wahl, A. 2004. European labor: the ideological legacy of the social pact. *Monthly Review* 55(8).

Waterman, P. 2001. Trade union internationalism in the age of Seattle. *Antipode* 33(3), 312–36.

Chapter 11

The Constitutive Inside: Contingency, Hegemony, and Labour's Spatial Fix

Jamie Doucette

Introduction: Between Labour Geographies

I often find myself doing two labour geographies that are implicitly related but whose relationality, at the theoretical level, remains difficult to articulate. One of these geographies is a more conventional political economy of labour: I try to find out where labour fits into a particular accumulation regime, or instance of capitalist social relations, using more conventional political economy categories of analysis, such as declining rates of profit, wage levels and organic composition of capital. The other geography is a political geography of labour: I examine how labour articulates itself as a political force, and attempt to determine where it fits into a particular hegemonic process or historic bloc. While I feel that these approaches complement each other, they can often conflict in terms of where the agency of labour is located. Is it to be found within the sphere of production? Or is it more likely to be located within that ensemble of relations between state and civil society which Gramsci (1971: 228–70) terms the 'integral state'? There are further locations one could add to these, such as the labour of social reproduction and the cultural habits of labourers. Furthermore, while political economy largely concentrates on class relations as the central antagonism within social space, Gramscian Labour Geography has focused on a wider distribution of power relations, such as the sets of relations organizing a historic bloc and the relations within this bloc and between state and civil society. Yet, to me, these different relations seem recursive and I attempt to hold them in tension with one another. This can be difficult, however, because the methodological distinction made between the relations of production and larger power relations among social identities is often taken for an ontological one. This methodological problem can be grasped by looking at the different ways that contingency, i.e. the unpredictable or indeterminate element of social relations, is understood in different approaches to hegemony and labour's spatial fix. This discussion should reveal some of the ways that Labour Geography, by embracing and expanding upon the understanding of contingency articulated by both approaches, provides some directions towards bridging the gap. Along the way, I will try to demonstrate some of this common ground by commenting on the politics of social cooperation in South Korea, where I have found myself attempting to understand the social cooperation process, and its failure to generate substantive agreements, in light of recent changes to the South Korean political economy and changes to the ruling hegemonic bloc under the reform governments of Kim Dae Jung (1998–2003) and Roh Moo Hyun (2003–08).

Social Cooperation, Economic Crisis, and the Reform Bloc

Understanding the social cooperation process in South Korea requires a perspective sensitive to changes within the political economy of Korea, particularly the restructuring of its developmental state model (see for instance Amsden 1989, Woo 1991, Woo-Cumings (ed.) 1999), as well as to changes within the integral state, i.e. political society plus civil society, brought about by democratization. Thus, there are two levels of influence that need to be addressed here. One involves the political economic model that Korea has followed, and the other involves the historic bloc of reform forces that has attempted to use social cooperation as a form of legitimation.

Similar to the fusion of political parties and labour movements in independence struggles in other developing countries (see for instance Jauch and Bergene's discussion of the Namibian trade union movement in this volume), the process of democratization in Korea was influenced by the fusing together of progressive and liberal forces into a united oppositional bloc. This historic bloc of reform forces finally came to power in the governments of Kim Dae Jung and Roh Moo Hyun. These reform governments attempted to create a new nexus between state and civil society, by expanding the role of reform intellectuals, NGOs, and the labour movement in policy making. I regard the labour movement here as an important part of the reform bloc and thus of civil society. Tripartite agreements were meant to be an important part of the increased participation of civil society within the state; however, they also resulted in a weakening of hegemony for the reform bloc, mainly because of the divisions they created when social partnership became a euphemism for neoliberal restructuring and, within the labour movement, a substitute for militant working-class politics (see for instance Gray 2008).

It would be wrong, however, to regard the tripartite process merely as a political project. The restructuring of Korean labour relations also has its roots in the growing profitability crisis among Korean firms that led to mandates of neoliberal restructuring by both the Kim Young Sam government (1993–98) and reform governments after it. Since the June Democratic Uprising and the Great Worker Struggle of 1987, wages had been increasing as a result of labour unrest (Jeong 2007). The restructuring of the industrial policies of the developmental state had also led to uncoordinated investment, speculative borrowing, and overcompetition in similar product lines (Shin and Chang 2003). The state had previously played a strong coordinating role in industrialization by subordinating finance to industrial capital, and when this was restructured it could not sustain the high levels of domestic industrial investment that it undertook as a developmental state. These factors combined and led to a decline in profitability, and, during the 1997 crisis, to a pressure by employers for concessions from labour and the government. Since Kim Dae Jung's government had favoured a shareholder model of corporate governance over industrial policy as its way out of the crisis, it also saw flexible labour markets as a potential solution to the crisis. This came in addition to socializing the debt of larger firms and selling them off to transnational capital and to the larger domestic conglomerates that survived the crisis. Here we see an intersection of two contingent events: a crisis brought about by declining profitability, resulting (partially) from the antagonism between wages and profit, as well as the restructuring of the older model, and a reform bloc supported by both workers and liberal reformers from the democracy movements pursuing neoliberal reform. This intersection led to efforts to create a new spatial fix for capital in the form of a national tripartite agreement facilitating the expansion of temporary and casual work. How, then, might the relation between these two different types of contingent events be better understood at a theoretical level?

Labour and the Spatial Fix

First, the spatial fix. This has been one of the more influential contributions of Marxist geography to the development of historical-geographical materialism. David Harvey ([1982] 2006) introduced the concept in his magisterial *Limits to Capital*, and he developed it to help describe the uneven geographical development of capitalism, and in particular the dynamics of over-accumulation through which capital produces space. According to Harvey, the roots of the spatial fix are found in Marx's thoughts on the 'inner dialectic' of capital as a social relation (Harvey 2001: 300): that is, the antagonism between wages and profit found in the relations of production and the concomitant tendencies of capital to move toward unbalanced accumulation and various crises, including crises brought about by technological development, keen competition, overproduction and speculation. In general, as capital accumulates, capitalists look for new avenues to stave off declining profitability, resulting in investments in technology, new product lines and the built environment. This process is itself a contingent one, and is based on a dialectic between fixity and fluidity. As such, it is never possible to completely predict solutions to accumulation crises. It is never certain where capital will find a new spatial fix for investment. Central to the process of finding a new spatial fix is the role of financial capital and the credit cycle, as capitalists try to offset crises by advancing capital in the form of credit to facilitate new investment. This dialectic speaks to the dynamic power of capital to shape the landscape in unpredictable ways. Conflicts between labour and capital can entail further investment in fixed capital or the geographic relocation of capital. Attempts to find a spatial fix are not merely limited to investments in new machinery, product lines or the built environment, but may include the financialization of capital and the expropriation of other capitalists through mergers, rent and speculation. Furthermore, spatial fixes to over-accumulation crises are not limited to capital accumulation through expanded reproduction but are also linked to other forms of 'accumulation by dispossession' (Harvey 2003) or extra-economic processes. While accumulation by dispossession, as primitive accumulation, is usually thought of as the creation of new wage workers through the dispossession of direct agricultural producers from the means of production, Harvey leaves the term open enough to include other forms of expropriation including war, imperialism and privatization. The essential point here, however, is that spatial fixes to accumulation crises can either work through an expansion of the dominant value-form of capital, as wage labour and investment, or it can work through the expropriation of resources or areas of social life that previously existed 'outside' of expanded reproduction. In fact, the two forms of accumulation are often inter-related and hard to disentangle.

Highly influential in geography and the social sciences, the concept of spatial fix has been essential in both the study of labour movements and of Labour Geography. In her *Forces of Labour: Worker's Movements and Globalization since 1870*, Beverly Silver (2003, see also Silver 2005) utilizes Harvey's concept of the spatial fix to chart the movement of both capital and worker unrest across the economic landscape. Silver shows how certain industries and product cycles, the automobile industry chiefly among them, tend to be haunted by what she calls 'Marx-type' industrial labour unrest wherever they migrate. This is worker unrest that puts pressure on firm profitability and causes capital to migrate or to pursue investment in innovation, increasing the organic composition of capital, in order to control labour costs. Silver finds that the complex production networks of automobile manufacture are sensitive to disruption because of sunk costs and the agglomeration of workers that such industries require. She shows how capitalists attempt to counter worker unrest by increasing automation (a technological fix) or advancing into newer spatial frontiers (a spatial fix) or new product lines (a product fix). While Silver reserves the term 'spatial fix' for capital migration it occurs to me that 'technological', 'product' and 'political' fixes are all deeply spatialized and should be regarded as spatial fixes in Harvey's sense of the term. Each fix she describes requires a reorganization of the landscape of production and a new spatial organization. Whether it be new machinery, product lines or bargaining procedures, each fix changes both the space and time of production, i.e. who works on what and where, as well as the organization of the workplace on the one hand, and hours of work, speed of production and delegation of tasks on the other.

The attempts made by Korean reform governments to create a tripartite process between labour, capital and the state can be regarded as a political fix to the problems of capital accumulation. As the old model was restructured, and wages increased from collective bargaining, Korean firms ran into a profitability crisis. During the 1997–98 financial crises, government intervention was required to reform the Korean economy, amidst pressure from international investors and domestic capital to find a solution to the crisis and a return to profitability among Korean firms. By offering the Korean Confederation of Democratic Trade

Unions (KCTU) full legal recognition in exchange for agreements on the use of temporary workers in multiple sectors, Korean capital attempted to create a new avenue for accumulation based on lowering wage costs and expanding irregular employment relationships. This agreement would also allow firms to contain union organizing by limiting the number of workers included in collective bargaining agreements and giving firms greater leverage to hire and fire casual workers. Though the 'Social Agreement for Overcoming the Economic Crisis' that the tripartite negotiations produced was later rejected by the KCTU's rank and file (see Table 11.1), the government accepted it as a de facto agreement and used it to legitimize Kim Dae Jung's other neoliberal reforms, such as bailing out the financial institutions that had rushed to lend short-term foreign credit to domestic firms and the selling off of bankrupt firms to both foreign and domestic investors. Using Harvey's terms, the political fix to the 1997 crisis involved accumulation by dispossession, through socialization of debt, fire sale of assets, and privatization, to eke out future profitability through expanded reproduction with lower wages and debt levels. One can see here how the concept of spatial fix is a useful analytic concept for understanding how crisis tendencies within capitalism lead to a reconfiguration of the economic landscape. However, fully grasping the contingency of this process, especially in terms of why a political fix at the national scale was sought instead of a different form of spatial fix, such as a product fix backed by government investment for example, requires a different understanding of spatial fixes that can better account for the political articulation of labour politics.

The conception of spatial fixes embraced by Harvey and Silver has at its centre an internal dialectic of class relations between labour and capital. However, the problem is that in Harvey's conceptualization of spatial fix the interaction between civil society and state involved in capital accumulation remains relatively underdeveloped (see for instance Jessop 2008: 178–96). Harvey's own approach to the state has largely been a derivative one, based on trying to understand which 'structures and functions within the state are "organic" to the capitalist mode of production and therefore basic to the survival of capitalist social formation and which are, in Gramsci's phrase, purely conjunctural?' (Harvey 2001 [1976]: 283). This distinction creates a problem for understanding the contingent relations which structure both the 'organic' moments of capital accumulation, i.e. the internal dynamics, and the larger social contexts in which capital accumulation takes place. For example, the Korean developmental state internalized much of the financial system during its developmental period. Viewed from the perspective of *laissez-faire* economics, and even some strands of Marxism, this would be an entirely inorganic role, but it would be wrong to assert that it was also merely conjunctural without positing some essentialist form of the capitalist financial system. While the dominant norm in capitalist social formations may be to valorize financial markets, and thus money as general equivalent, there seems to be a further degree of contingency shaping capitalist social relations that needs to be fleshed out.

Table 11.1 Tripartite negotiations in South Korea

1st Phase: Dec 1997–Feb 1998 Representatives of labour, capital and government participation	Social Agreement for Overcoming the Economic Crisis (rejected by union's rank and file, but recognized by government anyway)
2nd Phase: June 1998–Sept 1999 Business and labour groups repeatedly walked out of the process	No agreements reached
3rd Phase: Sept 1999–Dec 2008	2000: Agreement of Basic Principles on Reduction of Working Hours was reached
Incomplete representation of labour, including only the Federation of Korean Trade Unions (FKTU) and not the KCTU	2001: Agreement on additional five-year postponement of the introduction of multiple trade unions at enterprise level and elimination of payment to the full-time union officials by employers was reached
	2002 Agreement on Protection of Non-standard Workers, Employment Promotion of Young Generation and Privatization of Four Affiliated-companies of State Companies was reached
	2004: Resolution on the Break-up of Korea Electric Power Corp.'s Distribution Business and Agreement on the Revitalization of Employee Stock Ownership Scheme were reached
	2005: Agreement on Vocational Training for SMEs and Non-regular Workers was reached.
	2006: Agreement on promoting middle-aged and aged persons' participation in the labour market was reached.
	2006: Agreement related to measures to activate childcare services for low-income workers was reached.
	2006: Grand Tripartite Agreement on the Roadmap for Industrial Relations Reforms was reached.
(Other) Tripartite Representative Meetings 2004–2006 An Attempt by the Roh government to revitalize industrial relations and the KTC, participation by each party.	Two meetings were held in June and July of 2004 before the KCTU quit the process over the government's position on the Non-Regular Workers Protection Bill, but it returned in July 2006 in order attempt to negotiate the NRWP bill and other items. The bill was eventually passed without KCTU participation in the 'Grand Triparitite Agreement'

Sources: Im (2006), KTC website, http://www.lmg.go.kr/eng/index.asp.

Hegemony and the 'Constitutive Outside'

Gramsci's concepts, such as historic bloc, integral state and passive revolution, are useful for understanding attempts at social cooperation in South Korea and their failure. It was only through mass mobilization and the forging of a historic bloc that the democracy movements were able to defeat the previous developmental dictatorship, which accommodated this movement through a passive revolution, endorsing free elections and improving procedural transparency. In order to accommodate to this transition and the continued power of the old economic and political elite, oppositional groups remained fused together within a liberal-progressive bloc that supported reform-oriented politicians and political parties. They attempted to build a bridge between civil society and political society that could be used to reconfigure state policy. This was a counter-hegemonic project in opposition to the power relations that informed the authoritarian developmental states of Park Chung Hee and Chun Doo Hwan. These power relations cannot be reduced to relations of production, since others revolving around labour suppression, social regimentation, state-led industrialization and unequal gender relations were important. However, once in power, this reform bloc failed to become hegemonic as it became internally divided over the types of economic reforms it wanted to pursue. Some reformers were content to help institute a liberal market economy, while others preferred a redistributive social democracy. The dominant, liberal segment of the reform bloc under Kim Dae Jung won out and sought a compromise with capital. As such, labour groups and social democrats within the bloc found that their concerns were neglected and had failed to be hegemonized by the reform government. Choi Jang Jip (2005), a prominent reform intellectual and Chairman of the Presidential Policy Planning Committee under Kim Dae Jung, complains that though a plan for the parallel reform of the market and state administration was avowed, there was no attempt to use social cooperation mechanisms to effectively create an alternative system to neoliberal reform, and the plan was never concretized:

> Thus, everything became ambiguous. The scope and principle of a democratic government's intervention in the market was not defined. Thus, the only way to overcome the IMF crisis was to passively implement the reform package outlined by the International Monetary Fund. In regard to the question of how the market must be organized in a new environment called globalization, a model was not provided where the issues of chaebol restructuring, privatization, labour, employment, social welfare, etc. could be discussed within a single comprehensive framework. In the meantime, following the authoritarian development ideology, market efficiency and market fundamentalism began to gain power as a new hegemony. (Choi 2005: 190–91)

The failure to use the social cooperation process for progressive ends meant that the demands of labour became a significant blindspot among the democratic promises made by the reform bloc during and after the financial crisis. This had the effect of undermining the reform bloc from within due to factional conflict. However, the attempt at social cooperation did not end with Kim Dae Jung but carried on into the Roh government, who attempted to use tripartite meetings to legitimize neoliberal reforms to the labour market, in particular the Non Regular Workers Protection Bill that was aimed at expanding irregular employment relations. Without the full participation of the trade union movement, especially the KCTU, the flawed nature of the process slowly undercut Roh's ruling Uri party's legitimacy within the liberal-progressive bloc. Only the conservative Federation of Korean Trade Unions, a pro-government union from the dictatorship era, participated in the process while the KCTU sat most of it out. It would be difficult to grasp these tensions within the reform bloc without awareness of the contingency, the democratic struggle, that had brought both liberal and progressive forces together since the power relations involved in this struggle exceed the antagonisms that structure class relations at the point of production. Thus, an analysis such as Gramsci's is useful for understanding the breakdown of social cooperation and its precise location within a series of power relations not only between labour and capital, but also within the newly reconfigured terrain of the integral state and the liberal progressive bloc.

While Gramsci's understanding of hegemony and the historic bloc seems to complement Harvey and Silver's focus on the internal movements of capital, not all theorists of hegemony are so sanguine. For example, Ernesto Laclau (1990: 9) argues that the social antagonisms that structure capitalist social relations do not take place between labour and capital but between the relations of production and the social identities external to them. According to him, an internal dialectic such as the antagonism between wages and profit cannot account for transformations, since such an explanation would reduce social space to an internal movement, ignoring power relations deriving from other social identities. In Laclau's (1990: 17) phrasing the relations between these identities form a 'constitutive outside', 'an "outside" which blocks the identity of the "inside" (and is the prerequisite for its constitution at the same time)'. Laclau's understanding of hegemony blocks any sense of internal dynamics of capital accumulation and effectively undermines the over-accumulationist reading of spatial fixes as I have presented it thus far. The result is a theory of hegemony that embraces the contingency of social and political identities and is opposed to the necessity of a solution to the problems of profitability theorized by over-accumulation theorists. What, then, happens to the theory of spatial fix, if the 'outside' of the labour-capital dialectic is regarded as constitutive? It seems, from what I have discussed so far, that this dialectic is fatally undermined by such a reading. There seems to be little one can do to attempt to grasp a relationality between capital accumulation and other power relations, unless such a relationality is purely one of an exteriority that both constitutes and cancels out all interiority.

This might be a good time to recall that, for Marx, socially necessary labour time, and thus the tension between wages and profit, was a relative category. It results from a conjuncture between 'the average amount of skill of the workman, the state of science, and the degree of its practical application, the social organization of production, the extent and capabilities of the means of production, and by physical condition' (Marx cited in Harvey [1982] 2006: 16). Therefore, though the antagonism between surplus value and socially necessary labour is a constitutive tension for capitalist social relations, it is already contingent on external relations. While for Harvey, and other theorists of the spatial fix, it is this internal dialectic that is their main focus – and I have tried to show so far that it is the under-theorization of the political which causes problems – it is important here not to mistake the under-theorization of external forces for the determination of these relations by an internal movement. The problem Laclau raises then seems to be one of methodological focus rather than a clear ontological distinction between internal and external laws of motion. However, Laclau uses the contingent articulation of relations of production and 'outside' social relations, i.e. relations between different social identities, as an opportunity to privilege the discursive character of social identities rather than the recursive relations between social identities and relations of production. Instead of trying to introduce a greater reflexivity into political economy, Laclau instead argues that linguistic and rhetorical epistemologies are the necessary methods for grasping the logic of hegemony, as well as the nature of objectivity as such (Laclau 2004: 136–37, 151) since there is no necessary relation between social identity and the relations of production. Rather, it is the discursive relations between social identities, in particular, relations of equivalence, substitution and dislocation, involved in hegemonic articulation that constitutes objectivity. While this perspective reveals much about the discursive construction of power relations, it neglects entirely the materiality of relations of production, which are regarded only in terms of their discursive effects. Boucher (2008) argues that this is a critical error for Laclau, as it collapses a theory of a social formation into a discursive-linguistic theory (see for instance Hart 2002). The relationality between different regions of structures that might determine parts of a social formation, such as the relation between historic blocs and trajectories of capital accumulation that I discussed above, as well as the relation between materiality and discursivity, are obscured:

> [T]he postmarxian version of 'complexity' is a horizontal proliferation of hegemonic centres, which amounts to the multiplication of simple political antagonisms and not the complexity of an overdetermined social contradiction [...] While Laclau and Mouffe affirm the existence of the external world and the materiality of discourse, they claim that the being of every object is discursively constructed (Laclau, 1990: 97–134). This blocks the path to the regional distinction between social (discursive) practices and the materiality of the object (the natural properties of objects and extra-discursive conditions of emergence of discourse). (Boucher 2008: 95)

A Constitutive Inside?

Labour Geographers have previously tried to address tensions similar to those discussed by Boucher. Andrew Herod, for example, has reworked the concept of the spatial fix further to include the agency of labour to greater detail. In order to counter what he perceived to be a trend within geography to regard labour more as a factor of production than as an active historical-geographical agent, Herod borrows from Harvey's terminology to argue that workers are intrinsically involved in creating their own spatial fixes. 'Recognizing that workers may see their own self-reproduction as integrally tied to ensuring that the economic landscape is made in certain ways and not in others (as a landscape of employment rather than unemployment, for instance)', Herod argues, 'allows them to be incorporated into analyses of the location of economic activities and the production of space in a theoretically much more active manner than heretofore has been the case' (Herod 2001: 33–4). This perspective opens up the analysis of spatial fixes, affirming the tension between labour and capital but also supplementing it with an analysis of the multiple spatial dimensions that condition labour's agency in the process. Herod's own case studies show how labour movements produce a politics of scale through collective bargaining, as well as participate in the politics of urban boosterism and foreign policy. These investigations facilitate an approach to the politics of place and scale of production that is sensitive to the role of labour movements and the contingent sets of power relations that they act through. 'Even when [labour unions] are defeated in their goal,' Herod argues, 'the very fact of their social and geographical existence and struggle means they shape the process of producing space in ways not fully controlled by capital (2001: 17).' Therefore, for many Labour Geographers, the relation between capital accumulation and wider sets of power relations is already regarded as operating internally in theories of labour's spatial fix.

Another way of describing this relationality is to say that the concept of labour's spatial fix entails a productive tension between a 'labour theory of value', a concept useful for grasping tendencies toward overaccumulation, and a 'value theory of labour', a concept useful for grasping multiple sets of power relations conditioning labour. Elson (1979) first used the concept of a value theory of labour to explain that the value of labour in the production process was not simply the result of socially necessary labour time, but also power relations that strategically value and devalue the work of different social subjects. For instance, these power relations produced (often gendered) distinctions between social and private divisions of labour, and thus informed the spaces in which conflict between wages and profit might appear. However, this realization does not necessarily block the insights of the labour theory of value, but rather provides a necessary supplement that shows how the value of labour is determined from more than one source. In other words, it broadens the study of value to include relations between different social identities. In this way, the labour-capital antagonism can still be thought of as a formative and dynamic part of capitalism although it is here rendered relational to other power relations. They key point is that these two, the 'inside' and the 'outside', condition each other, but in such a way that also subverts any clear distinction, and one-way determination between them such as that embraced by Laclau. The inside of the capital-labour relationship already presupposes a complex work of valuation that does not necessarily block the internal movements of capital, but does reveal their contingency, or 'contingent necessity,' to borrow a phrase used by Jessop (2008). This framework of analysis is in contrast to Laclau's formulation in which any internal dynamic of capital is collapsed into the constitutive outside and made merely an effect of discursive power. Laclau's formulation eliminates many of the contradictions inherent in capitalist development from the analysis of hegemony because the 'outside' always subverts the 'inside'. The response from Labour Geographers, on the other hand, has been to see the two as mutually constitutive and not necessarily in antagonism. As Wright (2001: 560–61, see also Wright 2006) describes:

> I think a reading of Marx that emphasizes his view of materiality as in a constant state of production allows for an intersection with poststructuralist views of subjectivity to expand the concept of subversion to include the subversion of the discursive technologies so necessary to the devaluation of human beings. This reading of Marx utilizes his critique of value without circumscribing a vision of subversive agency to a strict allegiance to class politics, especially when there is little empirical evidence to support this approach.

Wright does this by showing how the value of labour power is not simply the result of its enrolment within the production process, but also of patriarchal social relations that allow labour to be devalued. This expands the spaces of Labour Geography from the point of production to other urban spaces where the subjectivity of labour and other social identities are devalued. Thus, the labour theory of value and the value theory of labour are held in tension here, forming a constitutive inside to capitalist relations, rather than one set of (discursive) power relations cancelling out the other (material set).

Conclusion

By way of conclusion, I would argue that this idea, of a constitutive inside to capitalist social relations, provides a conceptual solution to the methodological impasse between Marxist and post-structural approaches. It does this by holding the constituent parts of labour's spatial fix in tension, analysing capitalist accumulation without subsuming all power relations into an automatic labour-capital antagonism. Nor does it subsume all contingency into power relations purely 'outside' of capital accumulation. In my view, this necessitates a return to Gramsci's classical analyses of hegemony and historic bloc. As we have seen, Gramsci regarded the historic bloc as constituted by both the economy and politics as well as state and civil society, without necessarily reducing one to the other. In this way, the relation between the two different labour geographies of social cooperation I discussed in this essay can be analysed more clearly. The problems of accumulation – the declining profitability of Korean firms brought about by the decay of the developmental state model and by the ensuing labour upsurges – need to be read in tension with the mutation of the reform bloc. Once in power, reform forces within the state and civil society were not able to maintain their political unity. This was in many ways because their unity was always a contingent one, formed in opposition to authoritarianism and not able to transform itself once in power into a much broader articulation of democratic demands. Meanwhile, the embrace of *laissez-faire* by members of the reform bloc itself, combined with the pressure from capital to find a solution to the economic crisis, led to a hasty embrace of neoliberal restructuring, and an attempt to use tripartite means to legitimize it. This seriously undercut the ability of the reform bloc to forge a lasting hegemony and raised new tensions within it. Furthermore, the neoliberal orientation of reform governments also led to expanded conflict between the labour movement and the state. Meanwhile, the embrace of neoliberal reforms has only led to a slight recovery in profitability (Jeong 2007) that has since declined due to the current financial crisis. The current crisis undercut Korea's export performance and led to more calls for neoliberal reform. Understanding the contingencies that have led to this process, and the reforms and attempts at social cooperation that will be generated by it, will entail an analysis sensitive to both the political economic dynamics of capital accumulation and wider sets of power relations that inform them. I can only hope that the discussion of spatial fixes, hegemony and the constitutive inside that I have advanced in this short chapter might provide Labour Geographers with some useful theoretical suggestions for understanding the current crisis and the role of labour movements within it.

References

Amsden, A. 1989. *Asia's Next Giant: South Korea and Late Industrialization.* Oxford: Oxford University Press.

Boucher, G. 2008. *The Charmed Circle of Ideology: A Critique of Laclau and Mouffe, Butler and Zizek.* Melbourne, Australia: Re.press.

Choi, J.J. 2005. *Democracy after Democratization: The Korean Experience.* Seoul: Humanitas.

Elson, D. 1979. The value theory of labour, in *Value: The Representation of Labour in Capitalism*, edited by Elson, D. Atlantic Highlands, NJ: CSE Books.

Hart, G. 2002. *Disabling Globalization: Places of Power in Post-Apartheid South Africa.* San Francisco: University of California Press.

Harvey, D. 2001. *Spaces of Capital.* London: Routledge.

Harvey, D. 2003. *The New Imperialism.* Oxford: Oxford University Press.

Harvey, D. [1982] 2006. *Limits to Capital.* London: Blackwell.

Glassman, J. 2004. Transnational hegemony and US labor foreign policy: towards a Gramscian international labor geography. *Environment and Planning D: Society and Space* 22(4), 573–93.

Gramsci, A. 1971. *Selections from the Prison Notebooks*. Edited and translated by Q. Hoare and G. Smith. New York: International Publishers.

Gray, K. 2008. *Korean Workers and Neoliberal Globalization*. London: Routledge.

Herod, A. 2001. *Labor Geographies: Workers and the Landscapes of Capitalism*. New York: Guilford Press.

Jauch, H. and Bergene, A.C. 2010. Between revolutionary rhetoric and class compromise: trade unions and the state, in *Missing Links in Labour Geography*, edited by A.C. Bergene, S.B. Endresen, and H.M. Knutsen. Farnham: Ashgate, 127–140.

Jeong, S.J. 2007. Trend of Marxian ratios in Korea: 1970–2003, in *Marxist Perspectives on South Korea in the Global economy*, edited by M. Hart-Landsberg et al. Burlington, Vermont: Ashgate.

Jessop, R. 2008. *State Power: The Strategic-Relational Approach*. London: Sage

KCTU 2005. *KCTU Report on Recent Situation of Labour Laws and Industrial Relations for the Meeting with OECD Mission*. Available at: http://www.kctu.org.

KCTU 2007. *List of Workers Imprisoned Due to Trade Union Activities (As of January, 2007)*. Available at: http://www.kctu.org.

Laclau, E. 1990. *New Reflections on the Revolution of Our Time*. London: Verso.

Laclau, E. 2004. An ethics of militant engagement, in *Think Again: Alain Badiou and the Future of Philosophy*, edited by P. Hallward. London: Continuum.

Shin, J.-s. and Chang, H.-j. 2003. *Restructuring Korea Inc*. London: RoutledgeCurzon.

Silver, B. 2003. *Forces of Labor: Workers' Movements and Globalization since 1870*. New York: Cambridge University Press.

Silver, B. 2005. Labour upsurges: from Detroit to Ulsan and beyond. *Critical Sociology* 31, 439–51.

Woo, J.-e. 1991. *Race to the Swift: State and Finance in Korean Industrialization*. New York: Columbia University Press.

Woo-Cumings, M. (ed.), 1999. *The Developmental State*. Ithaca: Cornell University Press.

Wright, M. 2001. A manifesto against femicide. *Antipode* 33(3), 550–66.

Wright, M. 2006. *Disposable Women and Other Myths of Global Capitalism*. London: Routledge.

Chapter 12
Theoretical Approaches to Changing Labour Regimes in Transition Economies

Hege Merete Knutsen and Eva Hansson

Introduction

This chapter addresses the concept of labour regimes and what dimensions of the concept to focus on in explanations of labour activism in transition economies. The first part of the chapter starts with a review of conceptualizations of factory regimes and labour regimes. The concept of spatial embeddedness is central to the labour regime approach and important in analyses of power relations between state, capital and labour. It also addresses what a welfare-regime approach may have to offer studies of labour regimes, and how the struggle for labour rights is interlinked with the struggle for broader political rights. With examples from China and Vietnam, the second part of the chapter examines how labour regimes are formed in the context of transition economies and how this in turn affects labour activism.

Conceptual Framework

The Concept of Factory Regimes

Burawoy (1985) brings workers and their day to day experience to the forefront of analyses with his conceptualization of factory regimes. The point of departure is that factory regimes affect workers' struggles in different ways, but also that the struggles shape factory regimes. A factory regime is an overall political form of production which comprises the production apparatus and the political apparatus. The production apparatus deals directly with the labour process, and includes both relations in the organization of work and the relations of exploitation between labour and capital. The political apparatus deals with institutions that regulate and shape labour struggles, which in essence refers to how the state apparatus affects the production apparatus (Burawoy 1985). The micro-politics of the workplace and macro-politics of the state interact and shape the factory regime. Hence, factory regimes vary with 'both time and place according to the nature of the labour process, market forces, the reproduction of labour power, and the form of the state.' (Burawoy 1985: 111) The same labour process can be subject to different workplace politics, and development of factory regimes must be considered an interconnected international process.

Based on careful empirical studies at the factory level, Burawoy (1985) arrives at five main types of factory regime. These are: the company-state regime; the regime of market despotism; the regime of bureaucratic despotism of state socialism; the hegemonic regime; and the regime of hegemonic despotism. Of particular relevance to this chapter, the factory regime of bureaucratic despotism of state socialism is characterized by 'the coercive arm of the state at the point of production' and the notion that '[s]urvival outside work [depends] on adequate performance of work, monitored by party, trade union and management' (Burawoy 1985: 265). In the hegemonic regime, the vulnerability of workers *vis-à-vis* their employers is reduced by state labour legislation and social insurance. In the hegemonic-despotic regime, however, coercion prevails over consent as employers command workers' consent to forego rights and concessions in order to retain jobs. The transformations of factory regimes are historically contingent and may in some cases be linked to changes, and the struggle of labour, in factory regimes elsewhere.

From Factory Regimes to Labour Regimes

A regime depicts complex legal and organizational features that are systematically interwoven (Esping-Andersen 1990). The concepts of *local labour control regimes* by Jonas (1996, 2009) and *labour regimes* by Andrae and Beckman (1998) are perhaps the closest one gets to Burawoy's factory regime. Both concepts capture the micro-politics of the workplace and the macro-politics of the state, they refer to the complex of institutions, rules and practices that regulates relations between labour and capital both at the workplace and in society at large, and they are more scale-conscious than Burawoy (1985). Andrae and Beckman (1998) define reproduction of labour outside the workplace as the manner in which labour is trained and how it is supported by family, community and state outside the workplace, whereas Jonas (1996) makes a further distinction between politics of the living place and politics of consumption outside the workplace. According to Jonas (2009) labour control tends to stabilize around place-specific social practices which affect the social integration of labour inside the workplace. Moreover, local labour control regimes may 'infuse *other* workplaces and many aspects of local community and civic life throughout the locality or region.' (Jonas 2009: 3, emphasis added) In the same vein, Andrae and Beckman (1998) demonstrate how labour regimes can be studied in interlocking arenas such as the individual workplace, a given industry or sector, or at the level of the regional or national economy and politics. In contrast, Burawoy (1985) is primarily concerned with the fact that competition between places affects changes in factory regimes.

While one of Burawoy's (1985) objectives is to study factory regimes at the shop-floor level in order to arrive at a typology of regimes at a higher level of abstraction, Kelly (2002) studies how the nature of labour regimes affects workers at the shop-floor level and local-community level. This is a type of approach that requires analyses of how local labour regimes are influenced by social processes in other locations and at higher geographical scales, such as labour regimes at the local, micro-regional, national, macro-regional and global scale, as well as globalization processes more generally.

In transition economies, it is a case in point that migrant workers from the countryside often suffer harsher working conditions than local workers in the same factory (Rothenberg-Aalami 2004). Although Burawoy (1985) is concerned with how different categories of workers, such as young and old, respond differently to working conditions, he does not at the outset delve into the notion of different factory regimes for different categories of workers on the same shopfloor. In studies of how and why different categories of workers are subject to different labour regimes and working conditions, the concept of spatial embeddedness can serve as a useful tool. Kelly (2002) addresses the spatial embeddedness of labour control by discussing differences in interests and influence by different actors, such as local manufacturers, local and national authorities and foreign investors or supply chain managers. Interpreted this way, the concept of spatial embeddedness helps uncover where the various actors have their bases of power or gain additional strength.

As to whether the concepts of factory regimes or labour regimes can be applied in analyses of labour activism, we think that the concept of labour regime is more appropriate since it is more scale-conscious and since the need to move up and down geographical scales favours a semantically scale-neutral concept such as labour regime. It is also problematic that the concept of factory regime may unintentionally give wrong connotations that the regimes in question are uniform to all categories of workers on the same shopfloor.

Welfare Regimes and Commodification

A welfare state is a state that '[secures] some basic modicum of welfare for its citizens' (Esping-Andersen 1990: 18). This can be considered a minimum in order to qualify as a welfare state. From a more demanding angle, it is required that the daily routine activities of the state should be directed at the welfare needs of its citizens. The combination of inputs from the three institutions, markets, family and state, affects the nature and level of societal welfare. A welfare-state regime is the complex legal and organizational features that this combination of inputs results in. The point is that social rights are granted on the basis of citizenship rather than performance at work. This de-commodifies individuals in their relation to the market and strengthens the power of the worker *vis-à-vis* the employer. The concept of de-commodification implies that citizens have a right to services, and refers to 'the degree to which individuals, or families, can uphold a socially acceptable standard of living independently of market participation, [however] it is not an issue of all or nothing', and it does not completely eradicate labour as a commodity (Esping-Andersen 1990: 37). Commodification occurs when survival depends on the sale of labour power and markets are universal and hegemonic.

Welfare states and welfare-state regimes are concepts commonly applied to the historically specific condition of welfare capitalism. However, welfare is provided to varying degrees by the institutions of the state, market, family and voluntary sector in other types of society as well. Chan and Nørlund (1998) speak of *welfare socialism* in the state-socialist settings of China and Vietnam in the pre-reform period. Attention to welfare regimes is a way of addressing the sphere of reproduction of labour outside the workplace. Especially in the transition from state-socialism to increasing market orientation, it makes sense to study how provisions of welfare are shifted among the various institutions, how this in turn affects the nature and level of welfare in society, and the implications of these processes of commodification and decommodification of labour. This requires a concept of welfare regime which refers to the complex of legal and organizational features that characterize provisions of societal welfare, whatever combination of institutions that provide it and no matter how much the contribution. One may also distinguish between different welfare regimes of workers within a country. Both in Vietnam and China, migrant workers in areas of destination are often subject to welfare regimes that are different from what local labour is subject to (Rothenberg-Aalami 2004, Lee 1999, 2007, Hebel and Schucher 2006).

Political Regimes

Political regimes refer to 'the different institutional forms in which social conflict is organized, managed or ameliorated' (Jayasuria and Rodan 2007: 769). Political regimes are further identified by the mode of participation and contestation allowed, as well as by the relation between civil society and the state, or the mode of control exercised by the state over citizens and society. By limited or no competitive electoral structures, as well as strict control over what issues are allowed to be politicized, the political regimes of China and Vietnam do not encourage collective organization. The party-states of China and Vietnam secure individual compliance and collective order by political control, individualization of conflicts, repression and/or cooptation of emerging interests and social conflicts. The political regimes of China and Vietnam are thus characterized as authoritarian; nothwithstanding the changes that are taking place in modes of political participation and in how political control is being executed since the introduction of economic reforms. It is notable how technocratic rather than political and collective solutions are sought in institutional innovations of authoritarian states in order to open for more participation (Jayasuria and Rodan 2007) and to deal with the rising level of social conflicts. These changes typically continue a tradition of participation but without contestation. In China strategies of control differ with how important the social organizations are in providing public goods and to what extent they challenge state power (Kang and Han 2008). In Vietnam, opportunities of social participation have opened up, but these are accompanied by technocratic polices and rule by law that limits political contestation (Jayasuriya and Rodan 2007).

In most South-East Asian countries eradication, or brutal repression, of larger parts of the labour movements took place before industrialization gained speed. Although a dramatic surge in civil society organization pushed for democratization, many of the transitions from authoritarian rule that took place in the 1990s can mainly be seen as elite-negotiated. In general the resulting procedural democracies enjoy sufficient support from internal elites, but the *active* participation in politics by people's movements is still limited. These regimes are also exposed to limited pressure from international actors as they are treated as 'democracies' in international conversations. As the transitions took place in a specific neoliberal global conjuncture, the international focus has primarily been on the opening of markets and the protection of properties of international business. These are conditions favourable for authoritarian structures to survive and limit organization of civil society (Hewison 1999, Slater 2006). China and Vietnam embarked on a process of export-led industrialization in a context of neoliberal economic globalization without any independent labour movement.

Labour regimes and political regimes should be studied as interconnected processes because the nature of political regimes conditions the ways and means in which labour may respond to working conditions and general welfare, but also because labour may challenge and thereby incur changes upon political regimes. Struggle for specific rights at the workplace may in itself open political space and may translate into demands for wider political rights. Such a process may result in changes that benefit the demands of labour, but it could also result in counter measures that delimit the room of manoeuvre of any kind of social opposition.

Operationalization of Labour Regimes

Comprising the two spheres of the micro-politics of the workplace and the macro-politics of the state, the concept of labour regimes is highly comprehensive. Industrial relations and labour activism are but dimensions of the concept that interact with and are affected by what is going on along other dimensions. Transition economies are understood as economies in transition from state socialism to market orientation. With transition economies as the point of departure, explanations of changes in industrial relations and labour activism have to be explained with particular emphasis on the processes of commodification in these economies as well as increasing exposure to processes of globalization. Regarding the macro-politics of the state it is thus necessary to delve into commodification of welfare and how the state handles social conflicts that may arise from this. It is in the field of macro-politics that affect labour that the concepts of labour regimes and political regimes overlap. Politics of the workplace require focus on relations between the state-dominated and management-dominated unions, on the one hand, and the interests of different categories of workers, on the other. Both spheres are subject to international and global pressures and counter-pressures to remain competitive and assure 'expected' levels of working conditions and welfare of workers. This in turn requires attention to the spatial embeddedness of state, capital and labour which are likely to influence their respective actions and leverage. The point here is that in addition to the territorial embeddedness of the actors, they are embedded in relations with actors in other places and at other scales that affect their room of manoeuvre.

Labour Regimes and Forms of Labour Activism in China and Vietnam

Commodification

In pre-reform China the production system in urban areas was characterized by state-owned enterprises (SOEs) and work-unit (*danwei*) socialism. In addition, the household registration system (*hukou*) prevented people from rural areas to take up employment in urban areas. Still the migrant workers do not have citizenship rights to stay in the city, but those who have employment can be granted temporary permits of residence (Pun and Smith 2007). With the economic reforms that started in the 1980s, SOE workers who have been used to life-long employment face contract-based employment. Management is free to dismiss workers and recruits new and cheaper labour among the migrants from rural areas. Informal employment relations have surged, and remuneration based on piece-rates coupled with financial penalties in cases of rejects is adopted in labour-intensive production for the export market (Lee 1999, Hebel and Schucher 2006).

Welfare obligations of the SOEs are being replaced with insurance-based reforms in the areas of old age pensions, medical care, unemployment benefits, maternity benefits and housing. The system is based on contributions by the state, enterprises and employees, and it is administered by local labour insurance departments. However, it is difficult for impoverished SOEs to contribute to the funds, and the insurance programmes are unevenly enforced (Lee 1999). SOE workers in Vietnam share the same experiences. In 2007, less than 13 per cent of the workers were covered by a social insurance programme (interview in Hanoi with Vice Minister of Labour 2007), but an unemployment scheme based on insurance premiums was brought into effect on 1 January 2009.

Compared to the organized dependence of workers on the state and the production unit for work and welfare provisions in the *danwei* system, the commodification of the labour market and welfare provisions has taken place in the reform era (Lee 1999). Owing to the weaknesses of the insurance-based welfare system, means of reproduction outside the workplace become increasingly important to people's economic status and living conditions (Lee 1999, Hebel and Schucher 2006). These include contributions by other family members and kin, as well as access to diversified sources of income within the family.

Industrial Relations

The All China Federation of Trade Unions (ACFTU) is still the only legally recognized organization of workers in China. It is tied to the Communist Party and carries out its work under the leadership of the Party. The role of the ACFTU is both to assist the government in ensuring social stability, and to represent the demands of the workers so that inequalities and problems that arise do not lead to an independent worker movement. Moreover, the ACFTU is required by the government to educate workers on labour rights and encourage unions to establish in private firms. Unions shall be established in enterprises and government departments with 25 workers or more, but workers do not enjoy freedom of association nor the right to strike (Chan and Noerlund 1999, Hebel and Schucher 2006). However, strikes are not uncommon, though referred to as work stoppages and shut-downs. When workers elect local union officers they are generally nominated by the ACFTU. In private firms, managers are often union officials. Workers are sometimes not aware that unions exist in their workplace, or they are reluctant to join as the main concern of the management is to attract orders (Lee 2007, Global Labor Strategies 2008).

Similarly, in Vietnam the only legally recognized organization of workers is the Vietnam General Confederation of Labour (VGCL), but unions have to be established in all units employing ten workers or more. Unlike China, the law stipulates that leading trade union cadres are not required to be party members and that it is not part of the role of the union to participate in the management of the company. In practice, however, the roles of manager, trade union leader and party member can be performed by one person. In Vietnam workers in the foreign-invested sector have the right to strike, but procedures are cumbersome (Hansson 2001).

Labour Activism

Labour activism in China, ranging from collective petitions to riots, started to increase after 1989 and have surged since the mid-1990s (Lee 2007). In 2006, the number of cases reached 317,000 of which 14,000 were collective disputes (China Labour Bulletin 2007). There are large differences in how well the reformed SOEs do in the new market economy, and thus what they provide in terms of wages, pensions and other benefits such as housing and medical care. In addition, workers are organized in different work-units and there are differences in what levels of benefits different groups within the firm get depending on the contribution of the work unit to the insurance system. This results in intra-firm level protests (Lee 2007), but in 2005–06 there were also examples of protests that were coordinated across factories with the aid of the internet and cell phones (China Labour Bulletin 2007).

The number of labour disputes in Vietnam has also increased since the mid-1990s and accelerated the recent years. In 2007, the official number of strikes was 541 and involved an estimated 350,000 workers. In 2008, the number of strikes exceeded 700. Most of the labour disputes and strikes occur in the EPZs in the Ho Chi Minh City area, but a pattern of more geographically dispersed strikes that goes beyond the FDI sector is emerging.

In China and Vietnam, it is predominantly migrant workers who engage in collective actions, but laid-off workers and retirees from the reformed SOEs are also known to engage in such actions in China (Hebel and Schucher 2006, Lee 2007). Laid-off workers and retirees suffer problems in access to unemployment benefits and pensions, and without jobs they have less to lose from collective action than permanent and contractual workers. This corroborates the notion that attention to reproduction outside the workplace and access to welfare provisions by the state are important in explanations of activism.

Poor welfare provisions by the state in the countryside, poor working conditions in the cities, and missing opportunities of reproduction outside the workplace in the city, explain the particular squeeze that the migrant workers find themselves in. Pun and Smith (2007: 42) capture this in their concept of the dormitory labour regime which is understood as a 'hybrid outgrowth of global capitalism and legacies of state socialism'. Key characteristics of this labour regime are the low cost of labour, the low level of daily reproduction in dormitories, and close control of both production and daily reproduction by the management. Due to ample supplies of labour in the countryside reproduction of the next generation of workers is not an issue and wages are depressed. Labour is young, single and temporarily employed. Unlike for instance EPZs in other South-East Asian countries, this labour regime is considered systemic to China as it covers the majority of production workers and all industries irrespective of type of production, location and nature of capital. It rests on the urban-rural policies and residency control by the state, i.e. the legacy of the *hukou* system. It is also a legacy of the *danwei* system where urban resident workers were working and living together.

In China and Vietnam, collective actions arises in situations where migrant workers who suffer poor working conditions live together in dormitories. Interestingly, EPZs in Vietnam seem to be developing into a sort of 'schools of resistance' against the policies of liberalization. About 60 per cent of the workers in the EPZs are migrants from nearby or other provinces. In addition to the fact that they spend time together in the factory dormitories, they also have a broad web of contacts stretching far outside the EPZ compounds that they draw upon in their actions. Hence, close control by management by spatial confinement of labour is not always sufficient to prevent collective action. To what extent this can be attributed to the legacy of state socialism for example by norms and values or the degree of exploitation that the remnants of the *hukou* system open for, requires further examination.

In Vietnam, the VGCL (2006: 6) states that there are pressing and legitimate economic demands behind the collective labour disputes, and that strikes and actions are confined to industrial relations and are 'without any political motivation'. Nevertheless, the VGCL considers all strikes that have occurred so far as illegal because they have not followed the procedures of the Labour Law. In the 2005–06 strike it first appeared that the VGCL, in its eventual support of the workers main demand to raise the minimum wage, actually acted as representatives of the workers in negotiating with different levels of the bureaucracy (Tran 2007). However, the response was late and it was reported that the VGCL called in security forces. The workers gained their minimum-wage rise, but also ended up with a labour regime that seems to be moving in a more authoritarian direction and in which the party-state has chosen a more pro-business confrontational approach to strikes than before. This may also be exemplified by the new regulations that workers are liable for any economic damage incurred upon employers in case of an illegal strike. Deployment of security forces has not intimidated workers whose current strikes seem to be as resolute as before, and even militant, in pressuring the government for reforms.

In Vietnam the party-line, and in consequence the VGCL-line, has been to characterize the strikes as merely 'spontaneous reactions', denying them an organizational element which would threaten the monopoly of representation of workers by the VGCL and consequently the party-state. This illustrates how concerned the state is to maintain its political authority. The strike waves in Vietnam have shown that what starts as an economic bargaining process in the face of the unresponsiveness of the VGCL tends to develop into a political bargaining process that triggers counter measures by the state. This is similar to China where only actions that are not directed directly against the central state can be tolerated.

In China, Lee (2007) concludes that effective workers' protests have had positive social effects such as changes in pension reforms and punishment of abusive cadres, but that this has not led to more fundamental political change. Workers target local governments that set the local legal wage levels and are responsible for pensions and implementation of welfare reforms. They are more positive towards the central government because it is the central government that issues regulations that are favourable to workers.

Spatial Embeddedness

One consequence of the transition in China is that the growth of the private sector makes industrial relations more of a labour-management relationship as opposed to a state-labour relationship. From the angle of the state this 'depoliticizes' the demands of the workers and thus their protests may also seem less threatening to the central state (China Labour Bulletin, personal communication 2008). At the same time, however, economic growth has been accompanied by increasing levels of inequality and increasing social tension. As a result, in 2005–06 the government introduced the objective of attaining a harmonious society. This puts pressure on the ACFTU to focus more on how the restructuring of SOEs affect the workers; issues that are of particular importance to migrant workers; and to promote unions in the private sector. According to the new Labour Contract Law that became effective from January 2008, all workers shall be given a written contract and the contract becomes permanent after ten consecutive years of work. So far, intense protests against the new law by foreign business interests have not been accepted by the government.

Wal-Mart had to establish an enterprise union, against its initial will, when the ACFTU encouraged a sufficient number of workers to demand such a union. It has been argued that this may create precedence for workers in the private sector to establish more relevant unions of the bottom-up type, and that the new Labour Contract Law might open up a situation where unions become involved in genuine collective bargaining on behalf of their workers (China Labour Bulletin, personal communication 2008). Wal-Mart, however, outmanoeuvres 'grassroots unions' through top-down collective contract negotiations in order to assure the same contract in all of its stores in China (China Labour News Translations 2008a). In this case, and the case of OleWolff (Yantai) Electronics, the ACFTU at the local level supports the company management, while the ACFTU at the national level is more sympathetic towards the requirements of 'grassroots unions' (China Labour News Translations 2008b). This reflects the concern of local institutions about attaining local economic growth. Moreover, strong links between management and unions represent a mechanism for maintaining power by political control locally, which has its roots in the *danwei* system.

Labour NGOs with international and foreign links have proliferated in China since the 1990s and many of them are directed towards migrant workers. They empower workers by offering education, legal and psychological counselling. This in turn raises demands for legal justice and contributes to legal reform (Lee 2007). In contrast, international and foreign trade union organizations have been reluctant to have formal contacts with the ACFTU because it is controlled by the state and because of the Tiananmen incident in 1989. However, with the growing importance of China in the global economy, the International Trade Union Confederation (ITUC) has just announced that it is ready to begin a dialogue with the ACFTU (Global Labour Strategies 2008), and at OleWolff (mentioned above) the 'grassroots union' has obtained support for its actions from the Danish Confederation of Trade Unions and trade union organizations in Hong Kong.

Labour NGOs are not permitted to operate in Vietnam, but unlike China, Vietnamese trade unions may join and accept donations from organizations of international unions (Nørlund and Chan 1998, Hansson 2003). Foreign unions have stimulated a rise of occupation-based unions, but a greater openness to foreign and international trade union organizations has not been sufficient to significantly reduce the strength of international and foreign business interests in the Vietnamese labour arena. Pressure by the international business community contributed to a stricter regulatory framework on strikes after the minimum wage strike in 2005–06. This includes provisions that make strikers liable for material and economic damage caused to employers if they do not follow the legal procedure and authorize local authorities to force strikers back to work.[1] In the mid-1990s, when the first strikes in the foreign-invested sector occurred, open international interventions came mainly from foreign embassies, such as South Korea and Taiwan. These were a sort of 'scare tactics', commonly seen in other Southeast Asian contexts: if the government did not take care of 'the strike problem', investors would go elsewhere (Hansson 1995). During the 2005–06 minimum-wage strike, the tone had changed and in a concerted action both the American Chamber of Commerce, EuroCham and their Asian counterparts, publicly expressed deep concern that they had not been consulted in the raise of the minimum wage, including the timing of the raise and the wage negotiation procedures. They made direct threats to withdraw investments if the government did not ensure that the strikers returned to work (AmCham 2006, Fuller 2006).

On the one hand, the strikes have opened up political space as they have provided workers with more power to push through their interests from outside the formal framework of the party-state. On the other hand, regulations that have been pushed from outside by actors such as the chambers of commerce have tended to reinforce as well as further develop the authoritarian features of the system. Obviously, the party-state allows internationally organized interests of business to operate outside the party-state framework and within the national political space, while internationally organized interests of labour, such as international labour rights NGOs or advocacy groups, and even Vietnamese labour NGOs for that matter, are not permitted to operate in the national political space outside the formal relations of the party-state. These relations are effectively controlled by only permitting formal contacts with foreign trade unions exclusively under the oversight of the VGCL. In sum, the movement towards a more pro-business orientation of the VGCL and the MOLISA has resulted in confrontational and authoritarian measures which invoke further resistance among workers. This leads to genuine self-organization from below which in itself poses a challenge to the authoritarian party-state.

Conclusion

In studies of labour regimes and labour activism in transition economies we suggest that dimensions of the welfare-regime approach that affect workers directly and dimensions of the political-regime approach that focus on the phenomenon of the authoritarian state need to be included in any analytical framework. The welfare-regime approach facilitates an analysis of institutional changes that affect commodification and decommmodification of labour. It helps us come to grips with how welfare provisions in the state socialist contexts of China and Vietnam change with marketization, and how the responsibility of such provision is shifted from

1 Decree by the Ministry of Labour, Invalids and Social Affairs/MOLISA, 8 April 2007.

the state to the market and the family. This affects the nature social differentiation among workers which in turn affects the forms that activism takes. Migrant workers are squeezed in terms of welfare provisions, employment relations and working conditions. Together with laid-off workers from SOEs who have little to lose, they form the main group of protestors and engage in illegal strikes.

The political-regime approach helps us explain the close relationship between the ACFTU and the VGCL and their respective states, and the tendency of the state and unions to side with international business. The interests of business take precedence over labour when foreign business organizations put the state under pressure. At the moment, this appears to be somewhat more pronounced in Vietnam than China where foreign business has not managed to stop the new Labour Contract Law and the ACFTU has forced Wal-Mart to establish unions. However, Wal-Mart has regained leverage by enforcing top-down collective contract negotiations. In cases such as those involving the Wal-Mart and OleWolff, there appears to be a local-national dissonance whereby the national level ACFTU sympathizes with protesting workers in line with the objective of attaining a harmonious society, while the local level ACFTU supports foreign business in order to maintain power and attain economic growth at the local level.

Moreover, local-global relationships between local Chinese organizations and foreign and international labour NGOs have resulted in improved welfare for some workers at the local level. Relationships between local and foreign cum international labour organizations have not resulted in similar effects in Vietnam, as the former are not allowed to operate and the latter are under control by the VGCL and party-state. Whether this difference will translate into differences in power relations between workers and the state in the two countries and affect working conditions differently in the longer run remains to be seen.

References

AmCham Vietnam 2006. *Letter re: Minimum Wage and Illegal Strikes Involving Violence*. Available at: http://www.google.com/search?hl=en&lr=&q=site%3Aamchamvietnam.com+strikes%202006. [Accessed: 11.01.06].

Andrae, G. and Beckman, B. 1998. *Union Power in the Nigerian Textile Industry. Labour Regime and Adjustment*. Uppsala: The Nordic Africa Institute.

Burawoy, M. 1985. *The Politics of Production*. London: Verso.

Chan, A. and Nørlund, I. 1998. Vietnamese and Chinese labour regimes: on the road to divergence. *The China Journal* 40, 173–97.

China Labour Bulletin. 2007. *Speaking Out. The Workers Movement in China 2005–2006*. Available at: www.clb.org.hk.

China Labour News Translations. 2008a. *A Grassroots Union's struggle with Wal-Mart*. Available at: http://www.clntranslations.org/file_download/60.

China Labour News Translations. 2008b. *CLNT_OleWolff_intro*. Available at: http:/www.clntranslations.org/file_download/67.

Esping-Andersen, G. 1990. *The Three Worlds of Welfare Capitalism*. Cambridge: Polity Press.

Fuller, F. 2006. The workplace: strikers in Vietnam get little help from Europe. *International Herald Tribune*. Available at: http://www.iht.com/articles/2006/02/28/business/workcol01.php.

Global Labor Strategies. 2008. *Labor's Opening to China*. Available at: http://laborstrategies.blogs.com/global_labor_strategies_2008/03/labours-opening.html.

Hansson, E. 1995. *The Changing Role of Trade Unions in the Era of Economic Liberalization in Vietnam*. Politics of Development Group at Stockholm University (PODSU).

Hansson, E. 2001. Challenging authoritarianism: civil society and labour conflict in Vietnam, in *Civil Society and Authoritarianism in the Third World. A Conference Book*, edited by B. Beckman, E. Hansson, and A. Sjögren. PODSU, Stockholm University, 91–116.

Hansson, E. 2003. Authoritarian governance and labour: the VGCL and the party-state in economic renovation, in *Getting Organised in Vietnam. Moving in and Around the Socialist State*, edited by B. Kerkvliet. Institute of Southeast Asian Studies, Singapore, 153–85.

Hebel, J. and Schucher, G. 2006. *The Emergence of a New "Socialist" Market Labour Regime in China*. GIGA working papers. No. 39. German Institute of Global and Area Studies. Hamburg. Available at: www.giga-hamburg.de/workingpapers.

Hewison, K. 1999. Political space in Southeast Asia: 'Asian-style' and other democracies. *Democratization* 6(1), 224–45.

Jayasuriya, K. and Rodan, G. 2007. New trajectories for political regimes in Southeast Asia. *Democratization* 14(5), 767–72.

Jonas, A.E.G. 1996. Local labour control regimes: uneven development and the social regulation of production. *Regional Sudies* 30(4), 323–38.

Jonas, A.E.G. 2009. Labor control regime, in *International Encyclopedia of Human Geography*, edited by R. Kitchin and N. Thrift. Amsterdam: Elsevier, 1–7.

Kang, X. and Han, H. 2008. Graduated controls. The state-society relationship in contemporary China. *Modern China* 34(1), 36–55.

Kelly, P.F. 2002. Spaces of labour control: comparative perspectives from Southeast Asia. *Transactions of British Geographers* 27, 395–411.

Lee, C.K. 1998. The labor politics of market socialism: collective inaction and class experiences among state workers in Guangzhou. *Modern China* 24(1), 3–33.

Lee, C.K. 1999. From organized dependence to disorganized despotism: changing labour regimes in Chinese factories. *The China Quarterly* 157, 44–71.

Lee, C.K. 2007. Is labor a political force in China?, in *Grassroots Political Reform in Contemporary China*, edited by E.J. Perry and M. Goldman. Cambridge: Harvard University Press, 228–52.

Pun, G. and Smith, C. 2007. Putting transnational labour process in its place: the dormitory labour regime in post-socialist China. *Work, Employment and Society* 21(1), 27–45.

Rothenberg-Aalami, J. 2004. Coming full circle? Forging missing links along Nike's integrated production networks. *Global Networks* 4(4), 335–54.

Slater, D. 2006. The architecture of authoritarianism. Southeast Asia and the regeneration of democratization theory. *Taiwan Journal of Democracy* 2(2), 1–22.

Tran, A.N. 2007. The Third Sleeve: emerging labour newspapers and the response of the labour unions and the state to workers' resistance in Vietnam. *Labour Studies Journal,* 32(257), 257–79.

VGCL 2003. *Report of the Executive Committee of the Vietnam General Confederation of Labour (VII Tenure) at the IX National Congress of Vietnamese Trade Unions.* Hanoi: The IX National Congress of Vietnamese Trade Unions.

Chapter 13
Between Coercion and Consent: Understanding Post-Apartheid Workplace Regimes

Ola Anders Magnusson, Hege Merete Knutsen, and Sylvi B. Endresen

Introduction

In recent years Labour Geography has gained momentum as a sub-discipline in Human Geography. Labour Geography focuses on the agency of labour within capitalism and adds a scale and space dimension to traditional studies of labour relations. Inspired among others by the work on factory regimes by the sociologist Michael Burawoy (1985), Labour Geographers such as Jonas (1996, 2009), Andrae and Beckman (1998) and Kelly (2001, 2002) have developed conceptions of labour regimes in Human Geography. While Burawoy (1985) established ideal-type factory regimes that may be identified at any geographical scale and are also largely periodized, the approaches of Jonas (1996) and Kelly (2001, 2002) address the nature and dynamics of local labour regimes. Andrae and Beckman (1998) construct ideal-types but are also concerned with local dynamics. In other words, a variety of regime concepts are applied within labour studies. In addition to factory regimes, authors speak of labour regimes, labour control regimes, workplace regimes or work regimes. In the following *labour regime* is used in referring to all of them.

According to the conceptualization of Jonas (1996, 2009) and Kelly (2001, 2002), local labour control regimes develop under the concerted influence of local, national and global institutions and regulations. This chapter deals with links between the national context and workplace regimes at the local scale. The first objective is to discuss the links between the national and local scales in the geography of labour regulation and how this in turn affects local labour control regimes. Jonas (2009) argues that analyses of local labour control regimes represent a shift in explanatory focus away from the politics of *work*places to that of work*places*. Acknowledging that it is essential to examine power relations at the scale of local society in order to understand the dynamics of labour regimes, the second objective is to illustrate that it is important not to shift to the other extreme of omitting attention to *work*places in this endeavour.

A case study of the retail sector in Northern Namibia identifies different *workplace regimes* that have developed in the same industrial sector in one local society. Local society is here understood as *place* at the sub-national and sub-regional scale. It refers to the town of Oshakati, the administrative centre in Northern Namibia, located in the ruling Ovambo-elite's 'home'. Economic activities flourish in Oshakati where small and big black entrepreneurs, Chinese entrepreneurs and South African investors compete (Knutsen 2003, Endresen 2007). The case is instructive since differences between these businesses and the competition between them affect labour regimes and working conditions. In terms of power relations the three types of businesses are also differently embedded in Namibian society.

We aim by means of this qualitative empirical study to understand how labour-friendly work regimes are introduced and may take root, why repressive workplace regimes continue, and why *new* repressive regimes appear in Namibia, where labour legislation since Independence has favoured decent work relations. The chapter starts with a brief discussion of Gramsci's theory of hegemony and historic bloc that is used as a theoretical tool in analysing national labour relations and the links to national and local scales. Following an in-depth review of key conceptualizations of labour regimes in Labour Geography, we illustrate the theoretical points in the case study of Namibian retail. We attempt to stretch Gramsci's analysis of relations of force and hegemony across scale and extend it to the dynamic negotiation of workplace regimes.

Political Action as a Centaur

As already mentioned, this chapter rests on a Gramscian approach to the negotiation of political power, and since both Jauch and Bergene's (Chapter 10) and Doucette's (Chapter 11) contributions to this volume have dealt with this at length, we will here limit ourselves to a brief summary of his main concepts so as to lay the groundwork for their application in this chapter. Gramsci assumed a dual perspective on political action, and, according to Simon (1991), evoked the centaur as a symbol of its half 'human' and half 'animal' character. Central to Gramsci's perspective is the concept of hegemony, and, in his view, exercising political power is a delicate balancing act between persuasion and direction on the one hand and coercion and dominance on the other (Simon 1991). Hegemony thus becomes an alternative to force, and is a relation 'of consent by means of political and ideological leadership' (Simon 1991: 22). As noted by Doucette (Chapter 11), it is not very fruitful to seek explanations of social events with sole reference to the capital-labour relationship. In Gramsci's view (cited in Simon 1991) the analytical distinction between base and superstructure is not satisfactory since it carries with it the risk of misleadingly thinking that there is a clear-cut separation between the struggle against exploitation in the sphere of production and the political struggle for state power. Rather, in his view, 'the social relations of civil society interpenetrate with the relations of production' and he introduces the concept of historical bloc to 'indicate the way in which a hegemonic class combines the leadership of a block of social forces in civil society with its leadership in the sphere of production' (Simon 1991: 28). As noted above, central in the struggle for hegemony is the negotiation between direction and consent, on the one hand, and domination and coercion on the other. Hegemony 'presupposes that account be taken of the interests and the tendencies of the groups over which hegemony is to be exercised' thus necessitating certain compromises through which the hegemonic class must 'make sacrifices of an economic-corporate kind' (Gramsci 1932–34: 211). Another concept Gramsci introduces to capture these dynamics is passive revolution, which, as noted by Jauch and Bergene in this volume, is defined as the strategy chosen by the capitalist class whenever its hegemony is threatened (Simon 1991). The strategy involves extensive, top-down modifications in the social and economic structure in an effort to re-establish hegemony, and a main ingredient is reforms aimed at quenching the demands of opposing forces by granting them some benefits in a way which disorganizes the opponents and dampens their struggles down.

The Web of Regime Concepts

Burawoy (1985) employs the concepts of *factory regime* and *politics of production*. Factory regimes consist of a production apparatus, which refers to the organization of the relations of production and the relations between capital and labour at the scale of the workplace, and a political apparatus 'outside' the workplace which affects the production apparatus. The concept of politics of production is understood as an inseparable combination of economic, political and ideological aspects that shape factory regimes. Hence, national and local laws and regulations that affect production as well as national labour rights and national and local labour regulations that affect working conditions form part of the politics of production. Based on empirical studies, Burawoy (1985) establishes ideal-type factory regimes that in principle may be identified at any geographical scale. In addition, the categories of factory regimes that he identifies may be dominant in specific categories of countries at different points in time. His work also implies that one might speak of a tendency toward a globally dominant factory regime. Transformation from one category of factory regimes to another is a consequence of conflicts between workers and capitalists at workplaces; competition between companies; changes in social regulation that affect the spheres of production and labour; and the power relations between the Government, labour and capitalists in the struggle for hegemony.

Andrae and Beckman (1998) draw inspiration from Burawoy's work, using the concept of *labour regime*. Labour regime is understood as the regulation of relations between capital and labour. As with Burawoy's (1985) concept of factory regime, their concept of labour regime summarises the complex of institutions, rules and practices that regulate the relations between capital and labour as they manifest themselves at the workplace. They identify labour regimes at the factory level, generalize them and discuss how they apply in different groups of companies and in different locations. Andrae and Beckman's concept of labour regime thus includes both a workplace and a local/regional scale dimension.

At the national scale, Andrae and Beckman (1998) introduce the notion of a *corporatist labour regime*. In corporatist regimes, unions have, more or less, gained monopoly in the representation of workers in negotiations. Corporatist regimes can be divided into a state-centred and a more social form of 'corporatism'. In state-centred societies unions enter into agreements with Government which are mainly dictated by the latter. In the more socially shaped corporatist regimes, unions enter into agreements with Government based on their own strength. The first regime, which is relatively authoritarian, mainly occurs in fascist societies or in underdeveloped countries. The second regime mainly exists in capitalist welfare states (Andrae and Beckman 1998).

According to Burawoy (1985), transformation from one factory regime to another arises from changes; for example, in markets and technology, and how these changes are met by different agents in the form of acceptance or resistance. In times of such changes and conflicts, labour regimes may be put under pressure from both unions and companies, and working conditions may be re-regulated by the Government through new legislations. The conditions under which workers offer their labour power depend on their bargaining power, both at individual workplaces and in the labour market. This bargaining power is based upon a complex balance between social and political forces in society (Andrae and Beckman 1998). Thus, the ability to and success in organizing into unions are important instruments in order to strengthen the bargaining power. von Holdt (2005) introduces the concept of *apartheid work regime* in order to grasp the relations between capital and labour at workplaces in South Africa. In his conceptualization of work regime he addresses labour regulations, ethnic relations and workplace relations. As Burawoy (1985) and Andrae and Beckman (1998), von Holdt studies how relations between capital and labour in society at large affect relations between capital and labour at the workplace, and his concept of work regime includes both a workplace and a national dimension.

The concept of local labour control regime as applied by Jonas (1996, 2009) and Kelly (2001, 2004) distinguishes between the regulation of labour, on the one hand, and labour control on the other. '[L]abor regulation covers the wider institutions of state and economy, which nationally or globally govern wage labor and the workplace', while '[l]ocal labor control refers to the particular processes by which workers are integrated into the workplace and habituated to production and work in a given place or region' (Jonas 2009: 4). Hence, local labour control regimes are identified at the local scale, they reflect locally constructed practices, and they are formed by local conditions in combination with the influence of extra-local institutions and regulations. This largely parallels the distinction between the production apparatus and political apparatus in Burawoy's (1985) concept of factory regimes. In contrast to Burawoy (1985), Jonas (1996, 2009) and Kelly (2001, 2002) are more concerned with how labour regimes take shape and evolve locally, as well as their consequences to labour at the local scale, than generalizations into higher level ideal-types of labour regimes.

In local labour control regimes, 'local' refers to what is place-based or regional, while 'control' highlights a concern with 'the structures and constraints under which labour exists despite itself' (Jonas 2009: 4). Local labour control regimes can be both firm-specific and industry-wide. In addition, a dominant type of labour control regime can evolve at the local scale due to a tendency for labour control to stabilize around place-specific social practices. However, Jonas (2009) is sceptical to the notion of a tendency towards one single global labour control regime. This can result in overgeneralizations and, from a geographer's point of view, it is likely to obscure how place affects the nature of labour regimes and their consequences for labour.

In theorizing the nature of local labour control regimes, Kelly (2001) argues that it is important to pay attention to the nature and evolution of the state in each context, the historical development of labour regulation, how political power has been exercised at the local scale, discourses of political legitimacy and labour market practices, relationship between local power and national state power, cultural constructions of household relationships and gender roles, and agency of individuals. In other words, he contributes to the analysis of linking the extra-local to the local by an examination of power relations between actors at different geographical scales including national agencies, corporate investors, factory workers, industrial estate management companies, recruitment agencies, village/community leaders, municipal governments, provincial governments and labour organizations. In his study of local labour control regimes in the Philippines, he found that local regulatory regimes differ from the national ones as they are mediated through a range of institutions, and that the relations that constitute the local labour control regimes are informal and fluid.

The above literature review reveals that concepts of factory regimes, labour regimes, work regimes and labour control regimes all partly overlap in so far as they combine relations between capital and labour at the workplaces with local or/and national labour regulations in their conceptualization. However, the same term sometimes refers to different geographical scales and abstraction levels at the same time, which unnecessarily complicates the field of study. The approach of the local labour control regime stands out as the most context- and scale-sensitive of them, and Kelly (2001) demonstrates how power relations between key actors at different scales shape local labour control regimes.

Labour Regimes and Retailing in Northern Namibia: An Analytical Framework

The analysis of labour regimes in Oshakati, Northern Namibia, draws upon the contextual understanding in Knutsen (2003), Endresen (2007) and Jauch and Bergene (Chapter 10), and upon the approach to local labour control regimes, which is the most scale-sensitive of the regime concepts that have been outlined above. In doing so, there is a need of clarifying how the concept will be used at different geographical scales. In line with Jonas (2009), the concept of national labour regulation will be applied in addressing 'the wider institutions' at the national scale that in different ways interact with local conditions and form the local labour control regime. Formal and informal regulations, the organizational structures that these develop in tandem with, and ideology form part of the concept of institutions. As the concept of local labour control regime can be used to denote both firm-specific labour regimes and industry-wide labour regimes at the local scale, as well as a possible dominant form of labour control regime at the local scale, it is helpful to name the various micro-types of local labour control regimes differently. The analysis of labour regimes in Oshakati has different categories of workplaces as its study object. Hence, the local labour control regimes at this scale are referred to as workplace regimes.

There are two central aspects of *local* in the regulation of labour markets. The first aspect is that labour regulations are only local in their practice, that is extra-local regulation initiatives are being locally implemented. The second aspect is that extra-local regulations are being influenced by local regulation experiences which, in turn, are considered when designing new regulations (Jonas 2006, 2009, Peck 1996, Castree et al. 2004). The capacity of labour movements to take action at the national scale depends to a considerable degree on what kind of politics of labour and production that exist, and it also affects and is affected by labour relations at the workplaces. National labour regulation is shaped by the power relations between the agents involved, i.e. labour, capital and regulatory institutions. The power relations between the involved agents are in turn influenced by the dominating ideology (Castree et al. 2004) and by hegemony, the historic bloc and resistance between dominating and dominated classes. Furthermore, dominating discourses on labour issues may also influence power relations. Hence, due to the fact that labour regimes are based on a dynamic mix of social relations and power structures, labour regimes are constantly changing; and capitalist societies are characterized by a multitude of labour regimes. Labour regimes may differ between countries, within the same industry in the same country or even within a city. This leads to a great variation in workers' course of action at different workplaces (Jonas 1996, Castree et al. 2004), and we will analyse this by extending Gramsci's theorization of the negotiation of consent and coercion to the scale of the workplace. The case study of the retail sector in Oshakati thus demonstrates how power relations at this scale can be analysed as expressions of coercion and consent, which may complement Kelly's (2001) more actor-oriented approach (above) in explaining the leverage of agents.

In the retail sector, products for which there are ample supplies are prone to sharp competition on price. In developing countries such as Namibia, a number of small shops cater for the demand for cheap products among people with little money to spend on basic goods, and at the same time there is a large number of unemployed willing to accept jobs in spite of poor working conditions. Regarding strategies of competition, it might be assumed that strong competition on price affect working conditions negatively. Keeping working hours long, wages low and limiting social security are ways in with costs may be reduced. The skill level of shop assistants selling such products may be low, limiting their bargaining power even more. There are segments of the population with more money to spend who are looking for higher quality products. Shops that cater for these segments may combine higher prices for their products with poor working conditions. This is unless they require staff that is in scarce supply, their enterprises are closely controlled by the local government, or if it is part of their image to provide good working conditions. A good image in terms of high quality products, a nice shop interior and good working conditions can be a strategy to obtain and sustain a good footing in a new market for firms with sufficient capital to make such investments. The local embeddedness of the firms in question may also affect working conditions. Close relations between shop owners or investors and the local elite may result in authorities ignoring poor working conditions if this is deemed necessary for competitiveness and therefore conducive to local economic growth.

National Labour Regulation in Namibia

The main objective of *South West Africa People's Organisation's* (SWAPO) was to liberate the Namibian people from South African rule, although it started as a movement opposing the exploitation inherent in the capitalist economic system, its basis of recruitment being contract workers (Cronje and Cronje 1979, Jauch and Bergene in Chapter 10). As pointed out by Jauch and Bergene, the establishment of National Union of Namibian Workers (NUNW) is linked to SWAPO's development and struggle for independence, since the former developed out of the latter; although the relationship soon became an asymmetrical one of direction and control. For instance, SWAPO prevented NUNW from electing their own leaders, and thus continued to exercise control over the union. A SWAPO leader stated in 1981 that: 'a trade union in an independent Namibia would perhaps upset the economy and politics through strikes' (Bauer 1998: 60); while a union student noted: 'The purpose of trade unions in post-independent Namibia will be determined by the objectives of the state' (Bauer 1998: 59).

The historic bloc between SWAPO and the economic elite has resulted in a strengthening of capital as against labour, and SWAPO's development discourse rests on neoliberal policies such as privatization, economic growth, trickle down and the self-regulating market. Not only does capital have the upper hand in terms of economic power, they also enjoy cooperation and advantages from the Government in this historic bloc. The asymmetric relationship between capital and labour is further strengthened by NUNW's affiliation to SWAPO, which makes it part of the historic bloc. It provides SWAPO with a façade of ruling with NUNW's consent and thereby legitimizing its policies. The Namibian Labour Act, and especially its enforcement, therefore benefits capital and not labour. There are several tokens of this: no national minimum wage, no unemployment fund, no functioning pension fund and no medical benefit fund. Furthermore, the Government's use of the ideology of social partnership between capital and labour is an attempt at what Gramsci (1929–35) described as passive revolution (Jauch and Bergene in Chapter 10). As seen above, passive revolutions entail co-optation of alternative ideas until there ceases to be any substantive differences between them and the ideas that are hegemonic. The ideology of social partnership makes workers and their unions responsible agents in the economic development of Namibia and in ensuring an attractive market for FDI and TNCs. Hence, this policy further legitimizes the exploitation of workers. Consequently, the historic bloc between the economic and political elite, and their struggle for hegemony, has resulted in social partnership in order to manufacture consent. This has undermined the very principles of labour regulation, i.e. labour regulations are a means of preventing labour and capital from making unreasonable demands, since when the supposed neutral state takes the side of capital, the demands of capital are accepted while the demands of labour are not. Hence, in this context, capital can put forward unreasonable demands which labour is compelled to accept due to the hope that creating a capital-friendly environment will ensure new employment possibilities.

National labour regulation in Namibia fits the regulatory dimensions of what Andrae and Beckman (1998) refer to as a *state-centred corporatist regime* (above). The Labour Act seeks to promote good labour relations and fair employment practices, and sets out basic conditions of employment, or what may be termed minimum conditions. Wage workers have the right to organize themselves in and establish unions, and to bargain collectively. Laws against forced and child labour have also been introduced, as well as maximum working hours at forty-five hours per week with a maximum of overtime hours at three hours per day and ten hours per week. Workers now have the right to a fully paid annual leave and to paid sick leave at the same rate of payment as a normal working day. Furthermore, the right to strike is granted by law, although strikes can only be resorted to when conflicts concern industrial relations. In other words, workers cannot strike in solidarity or around political issues (LaRRI 2003).

Workplace Regimes in Oshakati

Interviews in 2007 with workers in retail shops in Oshakati form the backbone of the research discussed in this chapter. The shops represented four different categories of businesses based on an expectation that this may have implications for labour regimes. These were large international and South African chain stores, big national stores owned by white businessmen, big and small local shops owned by black businessmen and small Chinese-owned shops. The findings from the interviews were discussed with local and national unionists, and government officials were interviewed regarding regulation and enforcement of regulation (Magnusson 2008). The workplace regimes that are cast as ideal types in this study were identified in the shops under scrutiny. However, the size of our interview sample does not allow us to make *definitive* statements about the workplace regimes in each of the ideal types discussed, only to suggest tendencies. This warning is essential to keep in mind in the discussion that follows. Although the workplace regime of coercion was found in small Chinese retail outlets, we do not claim that what we found is the case in all 'China shops'. Likewise, we found decent work conditions in fashionable international chain stores, inspiring the ideal type workplace regime of consent, but this does not imply that we believe that all foreign chain stores have decent working conditions. Nor do we argue that workplace regimes are *static*: working conditions of the contemporarily decent modern chain store, run by white South-Africans with courses in human resource management, may toughen if the competitive situation deteriorates or the post-Apartheid searchlight is shut off. There is ample evidence in Oshakati retail of such falls from grace. In this study, categories of ownership and workplace regime are linked to owners' ethnicity. And the argument of dynamism is valid here too: we will argue that colour matters in Oshakati, but which colour is of *significance* at the workplace, shifts over time with politico-economic regime.

Three workplace regimes were identified; the coercive workplace regime, the workplace regime of consent and the workplace regime of abuse partnership. Where a *coercive workplace regime* prevailed the workers faced job insecurity, tough discipline, long hours, very low wages, few, if any, benefits, and intense control both during and after work. The relations to the manager were characterized by either distance or hostility and the managers were feared by the workers. The managers simply refused all requests for improvements and threatened workers if they complained. There were no negotiations or efforts to listen, compromise or ensure consent. The restriction of benefits and the extended working hours are in breach of the Labour Act. Workers who accept such hostile conditions lack alternatives.

Retail workers in the small 'China shops' reported harsh working conditions. They were deducted shortcomings in the till and forced to pay back through overtime work. In some instances the workers that slept in the shops had to be inside before ten o'clock or they faced punishment. Apart from the fact that it is illegal to live in shops, this is a violation of civil rights and resembles bonded labour and Apartheid working conditions. Unions found it difficult to organize these workers although they were neither pleased with their wages nor their working conditions. The Chinese owners tend to keep to themselves and not engage in local society, and they did not welcome unionization. The conditions in these shops seem to be accepted by authorities as a side effect of the larger investments that the Chinese bring into the country. Neither do local authorities bother much with working conditions in the shops owned by small black businessmen that also compete in the low-price market. A workplace regime of coercion was found here too. To understand why regulations are ignored, it is necessary to understand that small black businessmen are marginalized in Oshakati; they do not have much political leverage in terms of advancing protectionist policies (Knutsen 2003). But still, the employment opportunities of the poor are important in order to legitimize the economic policies of the elite. Strengthening of marginalized black entrepreneurs is government policy. Sometimes this involves turning a blind eye to violations of labour rights.

Where a *workplace regime of consent* prevails, the workers have high wages and most benefits. They can express disagreements and suggest improvements and they are represented by unions. Workers characterize relations between unions and managers as good. The Labour Act is complied with, and workers are pleased with wages and working conditions. Workplace regimes of consent were found in shops owned by international and South African businessmen. These are chain stores that have high prices and target high income customers. They claim to base their image on the workers' happiness, loyalty and pride. The rationale is that workers should feel responsible for the success of the shop, and therefore do a good job (Magnusson 2008, Endresen 2007). These multinationals bring labour standards from abroad; good working conditions are part of their competitive strategy and are not forced upon them by local authorities. Their definition of healthy management may, however, include extensive use of hi-tech surveillance, following workers' every move. Relentless competition among foreign investors endangers good working conditions. When new stores open up in Oshakati with their glittering façades, the old ones lose attraction.

The workplace regime of *abusive partnerships* is characterized by workers having low pay and working hours longer than allowed by law. They have few benefits but get paid for overtime, and management is negative to unions although they are permitted. The relations between management and workers are characterized by strong bonds of loyalty; labour relations are built on consent in patron-client relationships. Behind the façade of consent, power relations are asymmetrical and the workers frequently experience harassment. The labour regime of abusive partnership was found in shops owned by big businesses run by black entrepreneurs. At the time of Independence, they represented islands of modernity and the working conditions were much to be preferred to the Apartheid work regime. Black workers were even allowed at the tills. For some time now they have struggled because of competition from up-market foreign stores, trying to uphold their modern image. But they lost their up-market appeal long ago. A favoured solution is now franchising. Joining foreign chains, such as SPAR may, if this multinational follows its stated policy, mean the improvement of the working conditions of the retail workers (Endresen 2007). The shops of the big black entrepreneurs are not protected against competition from foreign capital, although their links with the political elite are well known. Their fall from grace during the past decade illustrates that competitive strategy may shift from modernization, i.e. technological improvements and attracting high income customers, to labour squeeze. As competition hardens, labour regimes may turn ugly.

Conclusion

In this chapter we have accounted for the many ways in which authors apply the concept of labour regime and related concepts such as factory regimes and work regimes. We agree with the notion of Jonas (2009) that national (and international) labour regulation represents the wider institutions that interact with local conditions and form local labour (control) regimes. In the same vein, we apply a context-sensitive understanding of labour regimes in which the point is not to generalize local types of labour regimes into national or global ideal types of labour regimes. Our analysis of local workplace regimes is inspired by Gramsci's theory of hegemony and historic bloc, where hegemony is understood as an oscillation between moments of consent and coercion. This theory helps us explaining how national labour regulation affects different workplace regimes at the local scale by highlighting power relations between capital, labour and the state and how these are characterized by different degrees of coercion and consent. As can be glimpsed, the pendulum swing between coercion and consent is dependent upon power relations prevailing both in society at large and at individual workplaces.

In this particular case it is not primarily differences in the local embeddedness of the shops that explain differences in the workplace regimes. The shops are not under pressure by the Government or other local institutions to comply with labour regulation. The differences are rather due to their conditions of competition and strategies of competition. Resulting from the policies of the historic bloc that presently characterizes Namibia, the objective is to attract foreign capital in order to attain economic growth. By the welding together of the party and the union into one national-popular force, unions became part of the establishment, and there is limited opposition to deterioration of working conditions caused by intensification of competition that the economic policy brings with it. This line of thinking and acting has not met with opposition locally, but then again Oshakti is the 'home' of the Namibian political elite, which may explain why local authorities hardly engage in matters pertaining to working conditions. Paying lip-service to 'Black Empowerment', the elite promote foreign investors, bringing them to the Oshakati doorstep in trade fairs. The alliance between the SWAPO elite and foreign capital may explain the negligence of local businesses, big and small, as well as the lacking implementation of the laws securing workers' rights. Therefore, the defensiveness of unions can be understood in terms of their close links to SWAPO elites (Jauch and Bergene in Chapter 10). But in this particular case, however, the difficulties involved in organizing retail workers in shops where there are just a handful of workers with close relationships to their managers should not be underestimated.

What may seem paradoxical is that the workplace regime that showed the closest match to the Labour Act was found in a South African retail multinational. This is not uncommon; multinationals may not dare to behave differently. Apartheid history should also be drawn upon when explaining this. Being under constant public scrutiny, mistreatment of workers would not be condoned in shops owned by white businessmen. Whatever their nationality, they are associated with the Namibian population of European decent that is busily distancing itself from its Apartheid past, constantly flagging their good intentions. In the shade of the searchlight on white big capital, oppression of workers by black businessmen and Chinese petty capitalists thrive. We believe that we have shed some light on why the Namibian elite condone the continued existence of coercive workplace regimes, and even allow new such regimes to appear. The Chinese capitalists, however, seem mysteriously exempt from enforcement of labour regulations. This should be further studied, perhaps in terms of 'politics of the belly' (Bayart 1993), or is it 'payback time' for the Namibian elite? Local labour regimes can neither be understood without a profound knowledge of the politico-economic history that shaped the local context; nor without the play of the contemporary Namibian power elite in their neoliberal playground.

References

Andrae, G. and Beckman, B. 1998. *Union Power in the Nigerian Textile Industry. Labour Regime and Adjustment*. Uppsala: Nordiska Afrikainstiutet.

Bauer, G. 1998. *Labour and Democracy in Namibia, 1971–1996*. Athens, Ohio: Ohio University Press.

Bayart, J-F. 1993. *The State in Africa: The Politics of the Belly*. New York: Longman.

Burawoy, M. 1985. *The Politics of Production. Factory Regimes Under Capitalism and Socialism*. London: Verso.

Castree, N., Coe, N., Ward, K. and Samers, M. 2004. *Spaces of Work. Global Capitalism and the Geographies of Labour*. London: Sage.

Cronje, G. and Cronje, S. 1979. *The Workers of Namibia*. London: International Defence and Aid Fund for Southern Africa.

Endresen, S.B. 2007. *Doors Wide Open, Eyes Wide Shut: Foreign Direct Investment in Retail in Namibia*. Paper to the Nordic Geographers Meeting: Meeting the waves of globalisation: local, regional and environmental responses, Bergen, Norway, 15–17 June.

Gramsci, A. 1929–1935. Prison writings, in *A Gramsci Reader. Selected Writings 1916–1935*, edited by D. Forgacs. London: Lawrence and Wishart.

Gramsci, A. 1932–34. Some theoretical and practical aspects of economism, in *A Gramsci Reader. Selected Writings 1916–1935*, edited by D. Forgacs. London: Lawrence and Wishart.

Jonas, A.E.G. 1996. Local labour control regimes: uneven development and the social regulation of production. *Regional Studies* 30(4), 323–38.

Jonas, A.E.G. 2009. Labor control regime, in *International Encyclopedia of Human Geography*, edited by A.E.G Jonas, R. Kitchin, and N. Thrift. Amsterdam: Elsevier, 59–65.

Kelly, P.F. 2001. The political economy of local labour control in the Philippines. *Economic Geography* 77(1), 1–22.

Kelly, P.F. 2002. Spaces of labour control: comparative perspectives from Southeast Asia. *Transactions of the Institute of British Geographers* 27(4), 395–411.

Knutsen, H.M. 2003. Black entrepreneurs, local embeddedness and regional economic development in Northern Namibia. *Journal of Modern African Studies* 41(4), 555–86.

Labour Resource and Research Institute. 2003. *Promoting Worker Rights and Labour Standards: The Case of Namibia.* Available at: http://www.larri.com.na/research/Labour%20Rights%20Report.doc [Accessed: 28.04.06].

Magnusson, O.A. 2008. *A Revolution Betrayed. How Namibian Class Relations are Played Out in the Work Regimes in the Retail Sector of Oshakati.* Master thesis in Human Geography, Department of Sociology and Human Geography, University of Oslo.

Peck, J. 1996. *Work-Place. The Social Regulation of Labor Markets.* London and New York: Guilford Press.

Simon, R. 1991. *Gramsci's Political Thought: An Introduction.* London: Lawrence and Wishart.

von Holdt, K. 2005. Political transition and the changing workplace order in a South African steelworks in *Beyond the Apartheid Workplace. Studies in Transition*, edited by K. von Holdt and E. Webster. Scottsville: University of Kwazulu-Natal Press.

PART IV
Labour and Strategies of Capital

Chapter 14
Erosion from Above, Erosion from Below: Labour, Value Chain Relegation and Manufacturing Sustainability

Michael Taylor and John Bryson

Introduction

The purpose of this chapter is to analyse the pressures of an increasingly complex and turbulent economic environment on the manufacturing sector and its enterprises in the West Midlands region of the UK, and to discuss the implications of this for labour, especially in relation to skills formation. The heart of the region is the Birmingham and Black Country Conurbation which extends east to include Coventry and northwest to include Stoke on Trent. The pressures that confront this longstanding and once dynamic regional economy include the imperatives of global competition, the distinctive pressures of a national and regional economy dominated by large firms, all too frequently inappropriate government policies, and growing labour market problems. Currently, all these problems are exacerbated by nearly two years of recession.

For some time now, it has been unfashionable to explore the changing role of manufacturing in so-called 'post-industrial', developed market economies such as that of the UK, where knowledge-based, service activities, especially those engaged in financial, business and professional services, have been assumed to be the route to sustainable future economic growth. Yet, manufacturing still underpins most developed market economies, and only recently have business and professional services come to rank alongside manufacturing as a principal source of the UK's export earnings. Recession now focuses attention on manufacturing as the jobs that remain in the sector are lost rapidly revealing the export earnings ramifications for these so-called post-industrial societies of which the UK has been said to be one. And, in a region like the West Midlands where manufacturing, and especially engineering firms are major employers, the severity of the current recession and rising unemployment levels highlight the vulnerability of what was once considered a regional powerhouse within the UK economy.

Now the West Midlands regional economy is seen to be in severe difficulty, with a compound of problems exacerbated by recession. First, the West Midlands has an enterprise deficit in the sense that the rate of new firm formation in the region is below the national rate. Second, the region's businesses are said to be non-innovative (OECD 2004), with innovation seen by government as the principal source of future economic growth. Third, the West Midlands has been recognized as having an enterprise finance gap, with small- and medium-sized enterprises (SMEs) finding it more difficult to raise finance in this region than is the case in other parts of the UK. Fourth, the region has an ageing manufacturing workforce with younger workers being less willing to work in what are often seen as dirty and unrewarding jobs. Fifth, there is a skills deficit in the region, as measured by reported shortages and numbers going through job training schemes compared with the numbers in other regions. Damning though this catalogue of problems might seem, it has not gone unchallenged. There is evidence of high value-added niche manufacturing strategies being adopted by some of the West Midland's SMEs as a way of meeting competition. Also, patent data do not support the assertion of the West Midlands being non-innovative, and it seems likely that skills training statistics understate the numbers of workers receiving traditional on-the-job training in the region. Nevertheless, these five problems are indicative of a region with significant constraints on its future economic prosperity.

The aim of this chapter is to develop an explanation of why and how the West Midlands region, once the manufacturing powerhouse of the UK economy, has shifted from 'growth region' to 'vulnerable region'. The argument developed here focuses on two sets of interrelated issues: economic globalization involving corporate change, managerialism, a new international division of labour and supply chain management, that combine in the West Midlands as processes of *erosion from above*; and labour market and skills issues that impact on manufacturing in the region's economy to create processes of *erosion from below*. The compound of processes bringing erosion from above has brought *value chain relegation* to the West Midlands fuelled by the political economic tragedy of the UK motor vehicle industry that has been played out since the 1960s. It is argued here, on West Midlands evidence, that the dynamics of manufacturing and the dynamics of labour markets are symbiotically interrelated. Each has the ability to enhance or retard growth in the other, and neither can be fully understood without reference to the other.

The argument of the chapter is developed in four sections. Following the introduction, the second section briefly introduces the West Midlands regional economy, highlighting its manufacturing core. The third section explores the restructuring associated with erosion from above that has occurred in recent decades producing a region dominated by a very small number of corporate market makers, focussed on the car industry, and vulnerable. Component manufacturers are seen as having been relegated in car industry value chains as local corporate market makers have failed. The section also explores the high value-added, niche manufacturing strategies that have been adopted by some smaller West Midlands firms to counter the problems of globalization and erosion from above. The examples of lock making and the jewellery trade are used to explore the problems of this strategy. The fourth section explores the processes of erosion from below that are intimately inter-related with the processes of erosion from above, and find expression in a contracting skill base and are threatening the future of manufacturing in the region. It highlights the problems associated with an ageing workforce, changing skill needs, public antipathy towards working in the sector, and, in particular, perverse government policies. The bleakness of the future for manufacturing in the region is summarized in the conclusion. Is this the death of a manufacturing and engineering cluster or the dawn of a new start?

The Economy of the West Midlands

The West Midlands is a complex region characterized by immense variety. It is centred on the Birmingham and Black Country conurbation with other major centres of manufacturing in North Staffordshire, in Stoke on Trent, and in Coventry. The region has a workforce of almost 2.6 million and its manufacturing economy is associated with traditional manufacturing, engineering and metal-based industries, dominated in the second half of the twentieth century by vehicle and car industries. Now, hi-tech industries and, increasingly, a diverse range of business and professional services are major contributors to the region's economy (Bryson and Taylor 2006).

The metal working and engineering industries of the West Midlands have long been recognized as a type-example of a linked enterprise structure (Florence 1948, 1961, West Midlands Group 1948) within which the close linkage of firms has been argued to generate *external economies of scale*, creating efficiency and competitiveness. This interpretation sits easily with the current ideas on 'clusters'. However, research on the region's economy in the 1970s and 1980s, offered an alternative interpretation. Being unable to identify the cost efficiencies recognized in theory, this strand of research argued that the region's linked enterprise structure owed more to behavioural processes that made business in the region 'easier' rather than more cost effective and 'cheaper' (Taylor 1973, 1975, 1978, Taylor and Thrift 1982, Taylor and Wood 1973, OECD 2004).

Although the region's economic structure is still dominated by manufacturing, this is under threat as the competitiveness of low-value added or labour-intensive manufacturing is undermined by processes of globalization, and the creation of value-added in other sectors of the region's economy outstrips that of manufacturing. As a consequence, over the last thirty years the region has experienced considerable economic restructuring and turbulence accompanied by the rise of unemployment linked to the closure, relocation and downsizing of manufacturing enterprises. At the same time, manufacturing's contribution to the region's GVA has declined from 29.3 per cent in 1997 to 21.0 per cent in 2002.

However, the manufacture of motor vehicles and other transport equipment still accounts for one in seven of the region's manufacturing employment, although this ratio continues to fall. In 2002, 19.2 per cent of total employment in the West Midlands was in manufacturing compared to a national average for Great Britain of 13.4 per cent, and those manufacturing jobs were mostly for men rather than women. The West Midlands is still the most industrialized region in the UK with significant manufacturing activity in car and automotive components, engineering, metal processing and manufacture, ceramics, brewing and the making of carpets as well as becoming increasingly well-known for specialist niche manufacturing. The region's manufacturing jobs are important within the UK economy because of the employment multiplier effects they bring with them. Research in the US by Bivens (2003) has shown that every 100 manufacturing jobs supports 2.91 jobs elsewhere in the economy, compared with 1.54 in business services and 0.88 in retail. Importantly for the West Midlands this work shows that some of the largest employment multipliers are found in automotive, aerospace and primary metals, and the manufacture of computer equipment and office machinery.

Erosion from Above: Restructuring, Value Chain Relegation, and Niche Production

Although elements of the linked enterprise system of the West Midlands show signs of continuation to the present, the system has also undergone radical restructuring, especially in the last three decades in the face of economic globalization pressures that, at the regional scale, have instigated and exacerbated processes of erosion from above. Economic globalization is a set of processes associated with corporate restructuring and the transformation from the 1970s onwards of 'multi-site' multinational corporations (MNCs) into transnational corporations (TNCs) within which separate global operations are managerially integrated. The Fordist corporation of the 1960s and 1970s was transformed in the 1980s when new sources of finance became available to fund their expansion. Those funds were US dollars held in European banks by OPEC countries that, as they forced up oil prices during the 1970s oil crises, were unwilling to leave the proceeds of their oil sales to the United States in US banks for fear of them being frozen. These 'dollars out of captivity' created the Eurodollar market, financial wholesaling and the London Interbank Offering Rate (LIBOR) that facilitated the mobilization of massive new investment funds. Loaned to countries, those funds triggered national debt crises, especially in Latin America. Loaned to corporations, they transformed the global economy. Principally, they triggered an on-going orgy of corporate mergers and acquisitions. They made corporations increasingly 'fiscal' (Taylor and Thrift 1986), and changed the very nature of corporate managerialism as it had first been outlined by Alfred Chandler (1966). In many corporations, the three pillars of managerialism that Chandler had identified, production, marketing and management, were reduced to just two; marketing and management (Taylor 1999). What was created was a new international division of labour (NIDL) with production shifted to low cost countries and control of marketing and management (including the control of technology and R&D) kept in the triad developed countries. Outsourcing and subcontracting became the order of the day creating large and anonymous contract manufacturers. Subcontractor accepted subordination in fragmented value chains, with major corporations holding the whip hand of power. And in this process of change the major corporations became, to all intents and purposes, stateless. Their goal is now to minimize their global tax burden, using the same tax haven states to which their bonus-fuelled executives also flee.

Restructuring and Value Chain Relegation

For a regional economy such as that of the West Midlands, the impact of these processes of economic globalization are corrosive. Beginning in the 1920s, Fordism and mass production had brought major changes to the West Midlands regional economy: the limited company replaced the family firm and the entrepreneur; mergers brought a loss of local control and corporate concentration; the scale of production increased as did the extent of vertical integration. Fordist mass production created boom conditions in the West Midlands through the 1950s and 1960s, but left the region's economy with significant vulnerabilities. The number of 'market makers' within the region was radically reduced, as indicated by the fact that 48 per cent of its manufacturing employment was in just 25 companies in 1977 (Spencer et al. 1986). That concentration, especially in the car industry, had significant implications. Many small, independent engineering firms were transformed into subcontract component manufacturers tied into the car industry. The corporate market makers, orchestrating value chains of component suppliers, became both powerful and vulnerable: powerful in the sense of being able to demand government assistance, but vulnerable to union pressure. The region's metal and engineering industries lost their resilience (Bryson et al. 1996).

The restructuring of manufacturing and engineering after 1970, heralded an era of mass customization rather than mass production, particularly in the car industry, that left West Midlands economy striving to cope with change: a weak player in the service economy; labelled in the 1990s as having non-innovative small firms; and having serious skills gaps (OECD 2004). Globalization brought intensified competition and a dramatic escalation in the speed of forced economic change. The West Midlands manufacturing economy began to be eroded from above. In developed market economies, a new significance is being ascribed to technology, expertise and design, coupled with speed of action and reaction. In this climate, the once dynamic West Midlands region is now seen as a region under stress: generating new firms at rates lower than the national norm (Deloitte & Touche 2002, AWM 2004); having small firms disadvantaged by a gap in finance for business loans up to £50,000 (NEF & Nicholson 2003); having a major skills gap that is exceptionally high in the western, Black Country section of the Birmingham and Black Country Conurbation; being non-innovative (OECD 2004); and a weak player in the service/knowledge economy.

With the ending of major vehicle assembly in the region, and especially with the demise of MG Rover at Longbridge and the closure of Peugeot in Coventry, the locally based value chains in this industry have been decapitated. The West Midlands components suppliers in this industry have been relegated in global value chains. Now, they have only foreign-owned corporations to supply. They, and the region within which they operate, are now peripheral to the companies (and the countries) that control the value chains within which they are embedded. Survey evidence from the West Midlands suggests that the firms in these chains that remain in the region are unwilling and, in the face of recession, unable to take on or absorb the sunk costs of training and upgrading workers' skills. Instead, they take on the skilled but ageing labour released by the closure of failing firms – a short-term expedient in response to the pressures of erosion from above that impacts on the processes of erosion from below explored in a later section of this chapter.

High Value-Added, Niche Manufacturing

The erosive and corrosive impact of economic globalization on a regional economy such as that of the West Midlands has also triggered survival strategies among smaller firms subordinated within value chains. Economic globalization enhances the role of technology, expertise and design in the creation of competitive advantage. Interview evidence collected between 2003 and 2007, from metal manufacturing and engineering firms (forging, pressing rolling and shaping) and lock making and jewellery making in the West Midlands suggests that technology-based, high value-added niche manufacturing strategies are being adopted in an effort by local firms to remain competitive. SMEs in niche metal manufacturing are focusing on quality and good customer service, using CAD CAM to produce small batches with a rapid turnaround. They also have wider markets, reducing their commercial vulnerability. These high-end niche metal manufacturers also have the potential to be the 'seedbed' from which a very different metal manufacturing sector might grow in the region. Two West Midlands manufacturing sectors illustrate the changes that are occurring to achieve these new roles: lock making and the jewellery trade.

Lock making in the West Midlands, based principally in Willenhall, shows the differential adoption of globalization and high value-added strategies among its constituent firms (Bryson et al. 2008). Large firms like Williams holdings and Assa-Abloy (controlling brands like Chubb, Yale and Union Locks) have sought to improve production processes by introducing standardized products and enhancing efficiency in their supply chains (Bryson and Rusten 2008), and this has involved them in rounds of merger and acquisition, and the outsourcing of production to low employment cost countries. However, while in one respect these companies are disengaging from the regional economy, they are, at the same time, maintaining skills-based specialist production in the region, and also research and design. But, there is survey evidence to suggest that these activities will remain in the West Midlands only as long as there is skilled labour to recruit that the large companies do not have to train themselves. In effect, they are mining the current skills base while avoiding the sunk costs of labour training: manufacturing firms have become, at least in terms of labour, extractive firms.

In contrast with the strategies adopted by large lock manufacturing firms in the West Midlands, smaller firms in the industry seek to remain competitive by adopting high value-added, niche manufacturing strategies. At the heart of this approach is the taking on of specialist orders for high value products that allow them to maintain or increase their revenues and net profitability. They steer clear of the mass production of standardized goods and competing on cost. Instead, they 'compete by design' in terms of expertise, specialization and functionality.

Unlike the large companies that simplify their product ranges, smaller niche manufacturers in the lock industry tend to retain large product ranges, but that range is frequently designed around a common lock platform – a common product base. This strategy allows them to cut costs in three ways: first, by using design to take work out of their products by reconfiguring existing components in new lock designs; second, by substituting expensive with less expensive materials; and third, de-skilling production so that semi-skilled labour can be used releasing skilled labour to work on higher value items.

What is very clear in the West Midlands lock industry is that the local embedding of firms in a socially based cluster, that was strong in this industry up to the closing decades of the last century, is now being eroded from above. Small firms are now fiercely competitive in their efforts to capitalize on the business opportunities created by the closure their competitors: to acquire clients, trained staff, and discounted equipment. Sourcing raw materials from local suppliers appears to be important, but it may just represent organizational inertia or an unwillingness or lack of awareness of alternative suppliers.

The same shift in strategy away from competing on price towards non-price based competition, and competing through design is also evident in the Birmingham's jewellery sector. Design in this context, however, has operated in a different way with firms specializing in either a single product or the making of customized products rather than through production-led or customer-led strategies. Birmingham's jewellery producers now concentrate on high value added, customized jewellery.

Birmingham has a long history of jewellery production but developed little or no reputational capital. It was London that developed a reputation for fashion and design associated with the tastes and fashions of the aristocracy (Sparke 1987), while Birmingham, along with the adjacent Black Country, developed an artisan-based, group-contracting system that designed and produced jewellery efficiently and cheaply but without an identifiable local brand or reputation (see the discussion in Taylor and Bryson 2006). Birmingham had superb artisans and craftsmen but the credit for their work frequently passed to those who retailed it. In fact, at the beginning of the twentieth century, Wright (1913) maintained that often the best and most beautiful work that appeared in the West End shop windows as 'London made' was, in reality output of Birmingham designers and workers and in this assertion there is more than a modicum of truth (Wright 1913: 379). Obviously, a strong retail brand sells more product, but it also allows retailers to shift seamlessly through an anonymous population of designers and makers.

In recent years, in the face of global competition, surviving firms in Birmingham's high value-added sector have focussed on market niches that have limited their exposure to competition from low-cost production locations. Part of this strategy has involved focussing on bespoke products. Small niche producers, and larger producers too, have also emphasized the importance of British design, and bespoke design does partially protect a firm from the negative impacts of copying. Indeed, place-based associations, combined with bespoke products, provide opportunities to create high value added products, even to the extent that one Birmingham jewellery design-maker now sells into the growing Chinese market. In this context, British design and production provides these products with a special type of place-based consumer cachet. A consumer is buying into a form of British identity and also engaging in conspicuous consumption based around company and country branding, and the storytelling through a range of communications media (Bryson and Taylor 2008).

Erosion from Below

Just as businesses in the West Midlands struggle to cope with the pressures of global competition there is, however, clear evidence that the high value-added niche manufacturing strategies being pursued by firms may offer only a short-term stay along a trajectory of manufacturing decline. Nationally within the UK, there is now a well-recognized dwindling supply of skilled labour in manufacturing (especially in engineering) and associated sectors that is eroding the production capacity of the economy. While global competition can be looked on as erosion from above, often orchestrated through the outsourcing strategies of corporate organizations as they become increasingly 'fiscal' and focussed on marketing, management, branding and design, the dwindling of the skilled labour supply is, in effect, erosion from below. Here it is argued that this latter form of erosion is driven by four processes:

- the changing demographics of the UK's population, especially the problems created by an ageing workforce;
- the changing skills needs of manufacturing and engineering as production methods change, sectoral emphases shift, and new technologies emerge;
- public attitudes to work and employment in manufacturing and engineering; and,
- ill-targeted government policies at all levels. At one extreme, national policies have favoured the service economy over manufacturing, while at the other extreme, local training programmes have been at odds with the realities of skills formation in small firms.

All the elements of this picture of erosion from below are appearing strongly across the West Midlands region where they appear to be interacting synergicly to accelerate decline.

From a demographic perspective, the workforce of the UK is ageing, as is that of the West Midlands region within it. Recently, EngineeringUK (2009) pointed out that between 2007 and 2017 the UK needs a net 587,000 people in manufacturing. But, in sectors such as the aerospace industry within manufacturing, 60 per cent of the workforce will retire in the next 20 years. In the West Midlands, nearly a third of manufacturing workers are over 50, and the average age of an engineer is 58 (Tift 2008). As the *Birmingham Post* (2008) remarked in this regional context, 'When an older worker retires you will no longer have the same conveyor belt of younger workers to replace them'. In the region's automotive industry there are major difficulties recruiting technical, skilled and semi-operative occupations (Donnelly et al. 2005), mirroring the situation in other engineering and metalworking sectors (Bryson and Taylor 2006).

Simultaneously, technological change in manufacturing and engineering is also changing the nature of labour demand. The future of engineering, in particular, is seen in the UK in 'green' technologies and nuclear power (EngineeringUK 2009). What is envisaged is high end production that competes on branding, customization and service rather than price and volume, all of which changes demand for labour and skills. At the same time, the nature and form of production in engineering and manufacturing in general is changing to achieve greater levels of productivity and to provide a better working environment to meet health and safety requirements. In the engineering sectors, this set of processes is seen as increasing demand for managers and senior officials in firms and also for people to fill professional, associate professional and technical occupations (EngineeringUK 2009). There are already shortages in these occupations and the highest levels of hard to fill vacancies, exacerbated by the shrinking numbers of Chartered Engineers and Incorporated Engineers in the UK as many are retiring or are on the verge of retiring. Whereas these occupations made up 32 per cent of the engineering workforce in 1987, EngineeringUK predicts that by 2017 that figure will have risen to 47 per cent. In a relative sense, there may be falling demand for people in skilled trades and elementary occupations in engineering in the UK but skills shortages are still serious in these occupations too. Drawing on the *National Employers Skills Survey (NESS) 2007* that was published in 2008, EngineeringUK (2009: 157) remarked that: 'the greatest area of skills shortage [in the engineering industries], accounting for just over two-fifths of SSVs [skills shortage vacancies], is in the skilled trades occupations. These include metalworking production and maintenance fitters, motor mechanics, auto engineers, electricians, electrical fitters and steel erectors'.

But, filling these vacancies is problematic, not least because of public perceptions of and challenging attitudes towards manufacturing and engineering. Tift (2008) has reported a corporate human resources director as saying that many, including teachers, have misconceptions about what manufacturing involves: 'It isn't dirt under the fingernails, heavy work it used to be. Many manufacturers utilize the latest technology and their factories are very pleasant environments'. EngineeringUK (2009) expand on this point reporting that only 12 per cent of 11–16 year olds claim to have any idea or knowledge of what engineers do. Among 16–24 year olds only 45 per cent consider engineering desirable as an occupation, and it is judged to be 'a dirty messy job' (EngineeringUK 2009: 41). Students, it would appear, would rather become teachers, actors, police officers, professional footballers, fashion designers, doctors, pop stars or singers. In the West Midlands, the Chief Executive of the Manufacturing Advisory Service (MAS) has pointed to the consequences of these public attitudes: 'We can't rely on overseas graduates to fill jobs, the UK must grow its own skilled workforce and this will require long-term planning and in-house training by firms. If the UK can't provide the skilled workforce to design and develop new and innovative products capable of competing on a global scale then the future is limited' (Tift 2008).

The last 20 years has seen manufacturing and engineering treated ambiguously by government policy. Hutton (2008) argues that in comparison with other developed economies, the high value-added manufacturing sector, in particular, has been insufficiently nurtured by the UK government. He reports a common complaint made to him by industry leaders that a range of issues, '...from the anti-industrial priorities of the financial system to cultural attitudes to engineering', are holding high-tech manufacturing back. Yet the UK is the world's sixth largest manufacturing economy (EngineeringUK 2009). The UK government has been jolted from complacency by financial crisis, retrenchment and recession. It now sees manufacturing and engineering as a source of economic sustainability and social well being. But, the structure overseeing skills formation and training that has been constructed in the past 6 years is a formidable growth industry in its own right and comprises layer upon layer of bureaucracy. The West Midlands Regional Skills Partnership established to co-ordinate this activity has 20 organizations as members and, in addition, includes the region's universities and further education (FE) colleges. Significantly it involves employers' interest groups but it also involves the Learning and Skills Council and private companies set up to deliver skills training. It also involves Advantage West Midlands, an economic policy generating organization in its own right. Ballooning bureaucracy coupled with large numbers of schemes and initiatives has led Matlay (2004) to see this as a partial cause of the low skills problem in UK industry, especially in small and medium sized enterprises (SMEs). He sees owners and managers as increasingly bewildered by the complexity of the current system, making them more likely to reject any advice they are offered. Martin and Matlay (2003) argue that skills training has to meet the 'shop floor' needs of small firms. In addition, Freel (2005) has argued on West Midlands evidence from small firms producing for niche markets that employees' skills are grounded in tacit knowledge gained over long periods of time. As a result, quick-fix training schemes to develop generic skills are completely out of kilter with the way training works in small West Midlands firms (see also Warhurst et al. 2004).

Interviews with West Midlands lock manufacturers amplify these training issues till further. They see existing government policies emphasizing skills for the knowledge economy rather than the manufacturing economy. Faced with this problem, one lock firm has recruited new school leavers through employee networks for on-the-job training because certificated FE college graduates do not have the necessary skills. Indeed, kinship networks have been the basis of employment in this sector and other West Midlands metal trades for decades, with up to three generations of the family having worked in the same firm. Now these kinship networks, identified by Greene et al. (2001, 2002) also appear to be failing.

The progressive failure of kinship networks in West Midlands manufacturing is also, in part, responsible for the falling rate of new firm formation in the region. As research in the 1970s showed (Taylor 1973), in the context of the West Midlands iron foundry industry, most new small firms were set up by people who had previously worked in other firms in the same sector. It can be argued that kinship networks among owners and kinship networks among the members of their workforces would make the social distance between business owners and their workers in their workplaces quite small, encouraging new firm formation. As such, failing kinship networks might stifle future entrepreneurship in the West Midlands.

To compound the issues surrounding ill-targeted skills training, interviews with West Midlands lock firms show that the direct impact of government skills policy can be perverse. The adoption of information and communications technology (ICT) and associated e-commerce practices to enhance production efficiency and productivity is a central plank of government policy in the UK, promoted at the regional scale by regional development agencies like Advantage West Midlands. Though potentially beneficial to industry, it can also have unintended consequences. One small firm surveyed in the region had been encouraged to introduce ICT into its business systems but found that access to the Internet enabled it to identify overseas manufacturers of inexpensive locks. In quick order, the firm stopped producing locks and became completely dependent on internet-enabled imports. As it commented, 'to be honest the manufacturing side of the business has taken more of a back-seat as we have concentrated on developing the website and importing products to sell through [it]. Being able to sell products other than locks has steadied the ship' (Interview, 2006). Now, a manufacturer has gone, and a distribution company has been created with a reduced regional employment multiplier.

Large firms compete for skilled engineering workers that are laid off when firms close, but they offer no training or workforce development. Their reaction is a reflection of the short-termism that affects British manufacturing which now seeks actively to avoid the sunk costs of labour training, especially in the form of apprenticeships. It means that the long-term future for many engineering and metal working firms will be constrained by an inability to replace skilled workers. The implication is that the specialization and customization strategy may be a short-term solution to foreign competition that will eventually be undermined by skill shortages.

Conclusion

The argument of this chapter is quite straightforward. In the face of global economic pressures, West Midlands manufacturing, made vulnerable by the problems of Fordist mass production, malleable government and strong unions has sought to remain competitive through the adoption of high value-added niche manufacturing strategies within a deregulated market environment and with compliant unions. However, the principal value chain of the region built around the car industry has been decapitated by the demise of the region's major car assemblers, relegating the region's parts suppliers within the globally orientated supply chains of this industry. The firms of the supply chain now live off the skills base of the region rather than contributing to its survival and regeneration. Now, the short-termism of business decision-making, coupled with an ageing manufacturing workforce, jobs that are unattractive to the younger generation, and inadequate skills training, means that firms' niche strategies are almost certainly doomed to fail having been eroded from below. Far from ameliorating and redressing the problems pushing the West Midlands region into decline, government policies are seen as inappropriate and off target, and only weakly connected with the way businesses actually work within the region.

It is striking that one major player in the West Midlands regional economy of the 1960s and 1970s is now missing – organized labour and strong trade unionism. Certainly Thatcher broke trade unionism in the UK, but why are workers still reluctant to join them in the West Midlands, as in the UK as a whole? Countering processes of erosion from below would seem like fertile ground for organized labour, but one which has not been well cultivated. Perhaps the answer to this conundrum is to be found in the complexity and the interweaving of the processes of globalization, corporate change and the unequal power relations within and between corporate organizations and smaller firms, set against a backdrop of deregulation, the market myth and now recession.

References

AWM 2004. Advantage West Midlands. *OECD Local Entrepreneurship Reviews: West Midlands, United Kingdom*. Available at: http://www.advantagewm.co.uk/downloads/oecd-final-full-report-aug-2004.pdf.

Birmingham Post 2008. *Older Workers could Solve the Skills Shortage*. 14 August. Available at: http://www.birminghampost.net/birmingham-business/.

Bivens, J. 2003. *Updated Employment Multipliers for the US Economy (2003)*. Washington: Economic Policy Institute.

Bryson, J.R., Daniels, P. and Henry, N. 1996. From widgets to where? A region in economic transition, in *Managing a Conurbation: Birmingham and Its Region*, edited by A.J. Gerrard and T.R. Slater. Birmingham: Brewin Books, 156–68.

Bryson, J.R. and Rusten, G. 2008. Transnational corporations and spatial divisions of 'service' expertise as a competitive strategy: the example of 3M and Boeing. *The Service Industries Journal* 28(3), 307–23.

Bryson, J.R. and Taylor, M. 2006. *The Functioning Economic Geography of the West Midlands Region*. Birmingham: West Midlands Regional Observatory.

Bryson, J.R. and Taylor, M. 2008. *Enterprise by 'Industrial' Design: Creativity and Competitiveness in the Birmingham (UK) Jewellery Quarter*. DIME (Dynamics of Institutions and Markets in Europe) Working Paper 47 on Intellectual Property Rights. Available at: http://www.dime-eu.org/files/active/0/WP47-IPR.pdf.

Bryson, J.R., Taylor, M. and Cooper, R. 2008. Competing by design, specialisation and customisation: manufacturing locks in the West Midlands (UK). *Geografiska Annaler*, Series B, Human Geography, 90(2), 173–76.

Deloitte & Touche 2002. *Access to Finance: Opportunities and Constraints for Business Development and Growth in the West Midlands*. Birmingham: Deloitte & Touche for Advantage West Midlands.

Donnelly, T., Barnes, S. and Morris, D. 2005. Restructuring the automotive industry in the English West Midlands. *Local Economy* 20(3), 249–65.

EngineeringUK 2009. *Engineering UK 2009/10*. London: EngineeringUK.

Florence, P. 1948. *Investment, Location and Size of Plant*. Cambridge: CUP.

Florence, P. 1961. *The Logic of British and American Industry*. London: Routledge and Kegan Paul.

Freel, M. 2005. Patterns of innovation and skills in small firms. *Technovation* 25, 123–34.

Greene, A., Ackers, P. and Black, J. 2001. Lost narratives? From paternalism to team-working in a lock manufacturing firm. *Economic and Industrial Democracy* 22(2), 211–37.

Greene, A., Ackers, P. and Black, J. 2002. Going against the historical grain: perspectives on gendered occupational identity and resistance to the breakdown of occupational segregation in two manufacturing firms. *Gender, Work and Occupations* 9(2), 266–85.

Hutton, W. 2008. The knowledge economy. *Prospect,* 151(October) Prospect Manufacturing Survey, 16–20.

Martin, L. and Matlay, H. 2003. Innovative use of the Internet in established small firms: the impact of knowledge management and organisational learning in accessing new opportunities. *The International Journal of Qualitative Market Research* 6(1), 18–26.

Matlay, H. 2004. Contemporary training issues in Britain: a small business perspective. *Journal of Small Business and Enterprise Development* 11(4), 504–13.

NEF & Nicholson 2003. *Analysis of the Need and Demand for Development Funding of CDFIs in the West Midlands*. Available at: http://www.advantagewm.co.uk/downloads.

OECD 2004. *OECD Local Entrepreneurship Reviews: West Midlands, United Kingdom,* Report undertaken by Advantage West Midlands as part of the OECD Leed programme. Available at: http://www.advantagewm.co.uk/downloads/oecd-final-full-report-aug-2004.pdf.

Sparke, P. 1987. *Design in Context*. London: Guild Publishing.

Spencer, K., Taylor, A., Smith, B., Mawson, J., Flynn, N. and Batley, R. 1986. *Crisis in The Industrial Heartlands: A Study of the West Midlands*. Oxford: Clarenden Press.

Taylor, M. 1973. Local linkage external economies and the ironfoundry industry of the West Midlands and East Lancashire conurbations. *Regional Studies* 7, 387–400.

Taylor, M. 1975. Organisation growth, spatial interaction and location decision-making. *Regional Studies* 9, 313–23.

Taylor, M. 1978. Linkage change and organisational growth: the case of the West Midlands ironfoundry industry. *Economic Geography* 54(4), 314–36.

Taylor, M. 1999. The dynamics of US managerialism and American corporations, in *The American Century: Consensus and Coercion in the Projection of American Power*, edited by D. Slater and P. Taylor. Oxford: Blackwell, 51–66.

Taylor, M. and Bryson, J. 2006. Guns, firms and contracts: the evolution of gun-making in Birmingham, in *Understanding the Firm: Spatial and Organisational Dimensions*, edited by M. Taylor and P. Oinas. Oxford: Oxford University Press, 61–84.

Taylor, M. and Thrift, N. 1982. Industrial linkage and the segmented economy: 1. some theoretical proposals. *Environment and Planning A* 14, 1601–13.

Taylor, M. and Thrift, N. 1986. Introduction: new theories of multinational corporation, in *Multinationals and the Restructuring of the World Economy*, edited by M. Taylor and N. Thrift. London: Croom Helm.

Taylor, M.J. and Wood, P. 1973. Industrial linkage and local agglomeration in the West Midlands. *Transaction of the Institute of British Geographers* 59, 127–54.

Tift, D. 2008. Skills shortage biggest problem facing West Midlands companies. *The Birmingham Post*, 24 April. Available at: http://www.birminghampost.net/birmingham-business/.

Warhurst, C., Grugulis, I. and Keep, E. 2004. *Skills that Matter: Critical Perspectives*. Hampshire: Palgrave Macmillan.

West Midland Group 1948. *Conurbation: A Planning Survey of Birmingham and the Black Country*. London: Architectural Press.

Wright, H. 1913. A general survey of the trades, in *A Handbook for Birmingham and the Neighbourhood*, edited by G.A. Auden. Birmingham: Cornish Brothers, 365–91.

Chapter 15
Globalization and the Reworking of Labour Market Segmentation in the Philippines

Niels Beerepoot

Introduction

In a recent editorial, *The Economist* was quite clear on how contemporary globalization generates winners and losers in the labour force. 'In the 21st century competition between firms and industries is becoming less important than competition between individual tasks within firms in different countries. Rather than affecting entire industries, or whole factories, global competition will affect individual jobs – skilled as much as unskilled' (*The Economist* 2007). Competition is no longer just felt at the level of the nation or the firm, with its effects trickling down to the level of its labour force, but the labour force is now itself directly susceptible to increasing competition both within its country and between countries. Contemporary globalization makes the world more interconnected but this interconnected world is being segmented in new ways (Krishna and Nederveen Pieterse 2008). This raises the question how the 'new competition' in the labour market will have its local outcome in various parts of the world, and how it will lead to a reworking of labour market segmentation. The co-existence of high wage and low wage sectors is the defining feature of labour market dualism, the generalization of which is labour market segmentation (Fields 2007). Labour market segmentation theory finds its origin in Western capitalist societies during the 1960s and 1970s (see for instance Doeringer and Piore 1971, Carnoy 1980). Since then various studies have identified how segmentation of labour takes place at various scales, such as the international, national and local, and through the working of concomitant processes of globalization (see for instance Peck 1996, Castree et al. 2004, Bauder 2006, Chapter 2 of this volume).

The Philippines is a good example of a country that is currently experiencing a number of simultaneous processes that rework labour market segmentation and lead to a repositioning of the country within the international division of labour. The country is experiencing a rapid decline of its labour intensive manufacturing sector, e.g. garments, shoes and furniture, which leads to massive numbers of displaced workers with limited formal education (see Scott 2005, Ris 2007, Beerepoot and Hernández-Agramonte 2009). The country has difficulty in competing with lower cost locations elsewhere in Southeast Asia, particularly China and Vietnam. At the same time, the country creates a similarly impressive number of jobs as recipient of off-shored service sector activities like call centres and business process outsourcing (Rodolfo 2006, Magtibay-Ramos et al. 2008). In recent years the Philippines has experienced a rapid expansion of this sector owing to its supply of young graduates with proficient English language skills. This chapter will provide an overview on how contemporary globalization leads to the reworking of labour segmentation at the national scale, the local scale and with regard to the position of the Philippines in the international division of labour. The chapter draws on ongoing research in the Philippines into the local outcomes of the changing international division of labour. Research reported here on the decline in labour intensive manufacturing was carried out in 2007 through interviews with 20 key informants[1], and that on the Philippine business process outsourcing (BPO) sector is based on existing academic literature and locally collected secondary data on the recent emergence of the offshore service sector in the country.

Section two will give an overview of labour segmentation theory. In section three the Philippines is positioned in a changing international division of labour. This receives further attention in section four and five where the decline of labour intensive manufacturing and the emergence of the call-centre industry as examples of reworking of segmentation in this country are discussed. Section six provides the concluding remarks.

Labour Market Segmentation Theory

A labour market often consists of various sub-markets with different sets of knowledge, skills and competencies necessary for productivity in the different segments (see for instance Doeringer and Piore 1971, Carnoy 1980, Droogleever Fortuijn 2003). Segmentation is a pervasive dimension of the purchase, sale and use of labour in capitalist societies (Castree et al. 2004). Labour market segmentation theory challenges neoclassical economic theory and human capital theory on the grounds that workers and jobs are not matched smoothly by a universal market mechanism (Bauder 2001). Emphasis on the variability of labour systems over time and between industries and places is one of the defining principles of segmentation theory (Peck 1996).

The various labour market segmentation theories hypothesize and try to establish that there are several types of jobs in the labour market, each with distinct criteria for hiring and advancement, supervisory procedures, working conditions and wage levels, and each with generally different groups who fill the jobs (Carnoy 1980). In both objective and subjective sense the particular employment workers find themselves in structures their working lives. Switching between, and sometimes even within, different segments then becomes quite difficult (Castree 2004). Doeringer and Piore (1971) distinguish two labour market segments: one

1 A detailed description can be found in Beerepoot and Hernández-Agramonte (2009).

characterized by better, permanent, well-paid jobs with career prospects, the 'primary segment', and the other having temporary, badly paid jobs without any career prospects, the 'secondary segment'. According to Doeringer and Piore (1971) a secondary labour market differs from a primary market in that it has low wages, poor additional benefits, poor working conditions, little additional training, a large number of temporary contracts, a high level of part-time work, few career opportunities, and a high labour turnover rate. Carnoy (1980) makes a distinction between a 'high-education' segment, a 'unionized segment' and a 'competitive segment' in labour markets. Similar to the secondary segment, the competitive segment in labour markets consists of a large and poorly educated labour force competing for jobs with low wages and unstable working conditions. Only those workers who have no other job opportunities, but who must still work to live, will accept employment in this segment (Carnoy 1980). These jobs have little or no provision for upward mobility. The unionized segment comprises jobs where unionization, or the threat of unionization, has secured for workers a set of structures which regularize the employment relationship and restrict competition among workers within the segment and from workers outside the segment (Carnoy 1980.). The commonality between the various segmentation theories is that labour is required to be flexible, with a core of highly skilled and dedicated staff complemented by a secondary peripheral workforce with much poorer conditions, lower wages and security, with poor career prospects (Atkinson and Meager cited in Danson 2005).

The two segmentation theories outlined above were formulated in a period when globalization was not as deeply ingrained in many countries as it is now and when unions were often more powerful. Early segmentation theories do not pay much attention to how segmentation changes over time in the face of increased national and international competition and technological change. Moreover, early segmentation analyses predate thoroughgoing consideration of globalization (Gray and Chapman 2004), and the mechanism of segmentation theory was based on the historical development of national labour market structures (Peck 1996). The national scale is not the only scale where a divide between segments takes place. Various authors have stressed how labour segmentation is also a gendered and locally constituted process deriving from unique intersections of labour demand and supply structures (see Peck 1996 for an overview). A local labour market can be considered as the geographically specific institutionalization of labour market structures, conventions and practices. Local labour markets represent the scale at which labour markets work on a daily basis, and at which some of the more significant reproduction functions are sited, and are among the scales at which labour markets must be understood (Peck 1996). Various processes of segmentation of labour, like gender, race and class, take place at the local scale, and the impact of processes like restructuring of production is experienced most profound within local labour markets.

The 'new international division of labour' thesis by Fröbel et al. (1980) can be interpreted as an international conceptualization of labour segmentation. Written at the time of major industrial transformations in Western Europe, the term was meant to illustrate the increasing subdivision of manufacturing processes into a number of partial operations at different industrial sites throughout the world. Brought about by the development of transport and communications technology and an increasing subdivision of labour, a world-wide industrial reserve army has come into existence, because all these potential workers can now compete 'successfully' on the world market with workers from traditional industrial countries (Fröbel et al. 1980). Two key processes that currently rework international labour segmentation are the international migration of labour and the increasing outsourcing and off-shoring of work. The flow of migrants into the cyclical, secondary segment of the labour market helps secure the jobs of non-migrants in the primary sector (Bauder 2006). This would sustain traditional segmentation of labour as migration stabilizes the labour market for non-migrants. Otherwise it can be viewed as driving a wedge between groups in the labour force with migrants and non-migrants pitted against each other. International service off-shoring, which is sometimes even referred to as 'the second global shift' (see Bryson 2007), would break up traditional segmentations of labour as new groups of workers would be exposed to increased international competition. Service off-shoring will not only affect routine work, but will also affect many formerly protected highly skilled and well compensated jobs, the large bulk of which are concentrated in developed countries (Freeman 2005). The new economy is associated with rising economic inequality both between and within countries (Perrons 2007). In the popular media Thomas Friedman (2006) classifies workers along lines of their vulnerability to international outsourcing. The segmentation hypothesis proposed by Friedman involves a combination of functional and spatial segmentation of labour. In his conceptualization, only smaller groups of workers can, owing to their specific expertise, e.g. accountants or brain surgeons, or their local embedding, e.g. bakers, plumbers or cleaners, secure their long-term employment and resist immediate threats of globalization, although the last group in particular is vulnerable to the entry of new migrant groups into the labour market. For the majority of workers, it becomes necessary to constantly re-invest in their knowledge and skills in order to secure their future employability in a globally competitive labour market.

The overview of literature in this section illustrated that workers are, simultaneously, part of labour markets that are segmented at international, national and local scales. Labour market segmentation does not just reflect human capital, the structure of the household division of labour, or employer discrimination, nor, on their own, the requirements of the labour process or the configuration of welfare regimes and industrial-relations systems, but it is a conjunctural, joint outcome of all of these (Peck 2003). Owing to the lack of unambiguous criteria for the drawing of segmentation lines in the labour market, researchers have great liberty in the way they construct labour market segments (Bispo 2007). Processes of segmentation are not universal unbending laws of economics, but should be understood as tendencies, the realization of which is particularly sensitive to spatial and historical context (Peck 1996). This reduces a possible universality of labour segmentation theory as an approach to the categorizing of jobs and structuring of labour markets. Multi-determined and historically produced social phenomena are rarely amenable to being crammed into quantitative empirical categories (Peck 2003). However, individual case studies within various country contexts can contribute to broader synthetic statements on the lines along which segmentation takes place. Critical realist accounts of the theory-empirical interface typically anticipate an iterative research process in which case-studies help to clarify conceptual understandings that, in turn, shape future case-study research (Angel 2002). In this context, studies in the global south help to identify the lines along which segmentation takes place within labour regimes where workers experience limited formal protection and representation. Additionally, workers are often highly vulnerable to international shifts in production, particularly in labour intensive manufacturing, that reduce possible long-term prospects for employment within certain labour segments.

Positioning the Philippines in the Changing International Division of Labour

The Philippines provides a good example of the problems and dilemmas faced by many industrial areas in less-developed parts of the world as globalization moves forward (see Lall 2000, Mani 2002, Beerepoot 2005, Scott 2006). The country has difficulty with the competitiveness of many of its traditional labour intensive industries while it has not been able to make the shift towards more advanced industrial production. This leads to massive displacement of workers with limited formal education (see Beerepoot and Hernández-Agramonte 2009). At the same time, the country creates a similarly impressive number of jobs in service sector activities like call centres and business process outsourcing. In recent years, the Philippines experienced a rapid expansion of this sector due to it supply of (young) graduates with proficient English language skills (see JETRO 2008).

Job-loss is typically associated with industries relocating to lower cost countries or closing down because they are out-competed (see e.g. Hudson 2000, 2005, Danson 2005, Felker 2003, Scott 2006). There is a well-established story of such job loss in Western Europe where the re-employment of displaced workers often turned out to be difficult. New forms of job loss in Western countries involve the outsourcing of tasks to other countries or labour replacement by recruiting from newly available pools of workers. Outsourcing becomes a process where individual tasks can be transferred to that part of the globe where it can be done the cheapest. This often involves silent processes of job loss in Western countries where jobs fade away, often through smaller reorganizations, in sectors that still thrive and in which job loss hardly receives press-coverage (see FNV 2005, WRR 2007)

The rise of offshore service activities and the fall of labour-intensive manufacturing is a clear illustration of traditional labour dynamics and modern forms of task outsourcing. Modern forms of task outsourcing are determined by the educational and language abilities of service workers located in low-cost locations (Bryson 2007). It is more constrained by language as well as cultural nearness, or the ability of foreign producers to relate to customers located in other countries (Bryson 2007). Both processes could sharpen the divide between the various segments in the labour market. Business services generally employ young, more highly educated workers, but have been criticized for providing them with narrow task descriptions, limited training and few career opportunities within this sector (see Carr and Chen 2004). Through their lack of education, displaced workers from labour intensive sectors do not have access to the newly created jobs in the offshored service sector.

Labour Intensive Manufacturing Decline in Philippines

A weak base of home-grown industries and dependence on externally controlled and owned branch plants makes the Philippines vulnerable to international economic cycles. A number of labour intensive sectors in the Philippines, like garments, shoes, furniture, gifts and toys, are currently confronted with an absolute decline in their export value (see for instance Scott 2005, 2006, Ris 2007, Beerepoot and Hernández-Agramonte 2009). In the garments sector, of the 900,000 formal garments workers registered in 1994, only 311,000 remained ten years later (ICFTU 2006). In the cottage-based shoe industry in Marikina, a suburb of Manila, an estimated 5,000 jobs have been lost during the same period (see Scott 2005, Ris 2007). These industries are important employment providers for people with limited formal education. This significant role is often underestimated when the decline of these industries is analysed.

Manufacturing decline in the Philippines is partly the outcome of the forces of globalization and the relocation of production to lower cost production sites while various local conditions are additional reasons why the country has difficulty catching up technologically with its more advanced neighbours. With sluggish FDI and export trends, the Philippines is a prime example of a country that is caught in what Felker (2003) calls a 'structural squeeze' between an ascendant China and more advanced countries like South Korea and Taiwan. Compared to these countries, the Philippines continues to be a far cry from a strong state that could competently intervene in markets and guide economic development without industrial policies being captured by rent-seekers (McKay 2006). Research has indicated that entrepreneurial strategies to deal with the decline in labour-intensive sectors involve strategic relocation away from Metro Manila, into the Philippine countryside, outsourcing to the informal economy in order to benefit from lower wage levels, or relocation to other Southeast Asian countries (see Beerepoot and Hernández-Agramonte 2009).

Employment within labour-intensive manufacturing is often based on the 'nimble fingers' of female workers. Additionally, workers must have job search skills, loyalty towards their peers and submissiveness towards employers in order to secure a position (Beerepoot 2005). Their skills have often not been developed in a way to secure longer-term formal employment or access to a wider range of possible jobs. Women, who were the early winners in the globalization process, are now beginning to lose out. They are losing ground in factory production, in terms of both jobs and work-related benefits, and are increasingly reliant on home-based work, which pays less, and/or on the most marginal of own account activities in the informal economy, such as vending (Carr and Chen 2004). Only anecdotal evidence exists relating to where displaced workers from labour intensive sectors in the Philippines ended up. The common perception is that the ever expanding informal economy easily absorbs most displaced workers (see Ris 2007, Beerepoot and Hernández-Agramonte 2009). Even more pressing than in Western countries, displaced workers, especially females, who have a limited education and who are beyond a certain age, which can be as low as 30 years old, have few opportunities for re-employment in the formal sector. Within labour intensive manufacturing, age forms an important criterion for access to formal employment. Older female workers are 'stuck' in the sector and within their location of displacement. Despite their extensive experience they are often forced to accept similar employment in an informal setting (see Beerepoot and Hernández-Agramonte 2009). There are few government programmes for retraining displaced workers and often they do not reach the neediest people (see Ris 2007). Compared to entrepreneurs, government and unions 'abandoning' the sector is not much an option for these women. The 'spatial fix' of these workers leaves them stuck in the location of their former employment, even when opportunities for alternative formal employment are limited.

The Philippine Business Process Outsourcing Industry

Within traditional conceptualizations, services are considered to be produced and consumed locally (Bryson 2007). With the digitization of information, it has become possible, and generally cost-effective, to transfer information processing work, both in manufacturing and in services, to offices and work units that are remote from main premises, within and across national boundaries (Carr and Chen 2004). The potential numbers of jobs that can be off-shored are at this moment unknown, but are likely to be enormous (Dossani and Kenney 2007). What were formerly non-offshoreable services may, through operational changes in business models, consumer preferences, and/or technology, be made amenable to offshoring (Dossani and Kenney 2007).

Call centre operations, which are the largest segment within off-shore services, require access to a large pool of flexible, low-cost labour (Belt and Richardson 2005). The labour market for call centre agents is often characterized as a 'secondary labour market' of insecure, low-paid jobs with limited career opportunities (see Dekker, De Grip and Heijke 2002).[2] For developing countries with capable

2 See Glucksmann (2004) for a configuration of call centre work.

workforces, services relocation offers enormous employment and entrepreneurial opportunities (Dossani and Kenney 2007). This type of global system replicates that of export-led manufacturing in that large numbers of young women are being recruited by emerging institutions of the digital economy, such as call centres, thereby providing new opportunities for inclusion, but on terms which discourage long-term, permanent contracts or unionization (Carr and Chen 2004).

Unlike the first global shift, the geographies of the second global shift are partially determined by a country's colonial heritage (Bryson 2007). Language (especially English language) skills are a major segmentation criterion from which developing countries can benefit and this has the potential to sharpen the divide between developing countries. The Philippines is a prime example of a country that has benefited from the international outsourcing of service activities. In recent years, the country's call-centre industry has undergone rapid expansion thanks to its pool of college graduates with a good command of English. For United States customers, the Americanized English of the Filipino call-centre workers is considered to make the country competitive with India. Around 300,000 people (with nearly as many men as women) are employed in this sector (see JETRO 2008). The workforce in this sector is just a fraction of the national labour force of 37 million workers. However, the sector is important for profiling the country as a modern service based economy. The vast majority of the BPO centres and jobs, both nearly 80 per cent of total, are located in Metro Manila (BPAP 2008). The concentration of the offshore-service sector in the main urban areas and the educational requirements imply that access to these jobs is highly segmented and (especially) the rural poor are excluded. The workers in this sector are sometimes considered as 'petit bourgeois' as they are generally young and well-earning but with limited connection with other labour segments.[3] At the same time, they benefit from the presence of an informal economy that provides low wage services. The back-office services that move from the developed world to emerging economies are supported by the low-wage services of the urban poor (Krishna and Nederveen Pieterse 2008).

Key activities in the Philippine BPO sector are customer care, legal and medical transcription, logistics and accountancy and software development and animation (Shameen 2006). The majority of the jobs in this sector, around 200,000 jobs, are in call centres (customer care) and mainly deliver lower value added services (see JETRO 2008). The rapid increase of jobs in the BPO-sector raises the question of whether this means the replication of the branch-plant syndrome or does it provide opportunities for longer-term competitiveness. The BPO sector is not a major stimulus in terms of economic interdependence (Magtibay-Ramos et al. 2008). So far, employment is based on a narrow job description and the limited acquisition of additional knowledge and skills or the knowledge and skills replicable in other professions. The opportunities for upward mobility for workers

3 Professor Rene Ofreneo, University of the Philippines-School of Labour and Industrial Relations, personal communication.

within their employment are often limited in this sector although these workers earn much more than many other people of their age and level of education. Employment careers within the call centre industry are generally short, only a few years, because of the pressing work schedules and necessary night shifts to serve the predominately US clientele. The physical demands of night time work leads to a deterioration in the individual's health and the inverted work schedule leads to an increasing sense of social alienation (Rodolfo 2006). Within the national division of labour a new labour segment is created, consisting of young, high income, earners with only selective and uncertain longer-term employment opportunities within their sector of employment.

Conclusion

This chapter has focused on a number of simultaneous processes that are leading to a reworking of labour market segmentation in the Philippines and a repositioning of the country in the international division of labour. Labour segmentation theory provides the framework to analyse the outcomes of the 'new competition' on the labour market for workers by identifying their long-term prospects for employment within particular segments and how the various segments on the labour market are connected with each other. Segmentation of labour takes place at different scales, such as the local, national and international, and according to different criteria, e.g. age, education and skills. At the national scale segmentation takes place between the urban economy of Metro Manila, which is the country's outpost in globalization and where most new employment is generated, and the rest of the country. Within the urban economy of Metro Manila, the divide is created between those people with access to the new service sector jobs and those who are confronted by the decline of labour intensive industries, with the latter group often ending up in informal economy activities. In this sense, a traditionally important secondary labour market segment declines while a new segment is created. At the international scale, the Philippines belongs to the select group of developing countries that has foreign language skills and is able to benefit from the international offshoring of service activities.

The integration within the global economy of various labour segments in the Philippines is uneven and often temporary, and this becomes clear when they are seen in temporal and historic context. Labour intensive manufacturing was very important as an employment provider in the 1980s, providing jobs for women who were excess to the needs of the agricultural sector. The BPO-sector currently provides employment for a large number of young college graduates for whom there are few other jobs are available nationally. But, these jobs are part of a branch-plant economy whose longer term prospects are unclear. Both sectors explored in this chapter reflect the 'old' and 'new' employability skills that are required of workers for them to compete in increasingly flexible international labour markets. Labour intensive manufacturing is based on the presence of 'nimble fingers' that are available throughout the developing world. The BPO-sector demands language and social skills from workers as well as imposed 'cultural nearness'. The current shift towards non-manual knowledge work, requiring higher level skills and qualifications, leaves those without the ability to adapt to these changes no other choice than informal, low-paid and unstable work. At local scale, this creates a divide between those with access to newly created jobs and those whose role is to provide them with low-paid informal economy services. This chapter has emphasized that for the study of labour segmentation it is necessary to focus on how segmentation at different scales has its outcomes in particular sites. Current segmentation analyses still mainly focus on only one level of segmentation while this chapter emphasizes the necessity to identify the interrelation between the local, national and international scales. Socio-economic processes at the global scale often impose fundamental limitations on seemingly local issues (Lier 2007). Understanding the terms of hierarchical integration in the global economy of different labour market segments, and how globalization sharpens divides within local labour markets, demands a holistic approach to connect the different scales at which labour segmentation exists.

References

Angel, D. 2002. Studying global economic change. *Economic Geography* 78, 253–55.

Bauder, H. 2001. Culture in the labour market: segmentation theory and perspectives of place. *Progress in Human Geography* 25(1), 37–52.

Bauder, H. 2006. *Labour Movement: How Migration Regulates Labour Markets.* Oxford: Oxford University Press.

Beerepoot, N. 2005. *Collective Learning in Small Enterprise Clusters: Skilled Workers, Labour Market Dynamics and Knowledge Accumulation in the Philippine Furniture Industry*. Ph.D. Dissertation, University of Amsterdam.

Beerepoot, N. and Hernández-Agramonte, J. 2009. Post MFA-adjustment of the Philippine garments sector: women's cooperatives amidst manufacturing decline. *European Journal of Development Research* 20(3), 362–376.

Belt, V. and Richardson, R. 2005. Social labour, employability and social exclusion: pre-employment training for call-centre work. *Urban Studies* 42(2), 257–70.

Bispo, A. 2007. *Labour Market Segmentation: An Investigation into the Dutch Hospitality Industry*. Rotterdam: ERIM Ph.D. Series Research in Management, 108.

BPAP 2008. Creative recruitments to sustain BPO growth. *Newsletter Business Processing Association Philippines* 2, 17.

Bryson, J. 2007. The second global shift: the offshoring or global sourcing of corporate services and the rise of distanciated emotional labour. *Geografiska Annaler* 89B, 31–43.

Carnoy, M. 1980. Segmented labour markets: a review of the theoretical and empirical literature and its implications for educational planning, in *Education, Work and Employment II*, edited by M. Carnoy, H. Levin, and K. King. Paris: UNESCO/IIEP.

Carr, M. and Chen, M. 2004. *Globalization, Social Exclusion and Work: With Special Reference to Informal Employment and Gender.* Geneva: International Labour Office. Available at: http://www.ilo.org/dyn/dwresources/docs/625/F1146925582/gender%20and%20globalisation.pdf.

Castree, N., Coe, N., Ward, K. and Samers, M. 2004. *Spaces of Work: Global Capitalism and the Geographies of Labour.* London: Sage Publications.

Danson, M. 2005. Old industrial regions and employability. *Urban Studies* 42(2), 285–300.

Dekker, R., De Grip, A. and Heijke, J.A.M. 2002. The effects of training and over-education on career mobility in a segmented labour market. *International Journal of Manpower* 23, 106–25.

Doeringer, P. and Piore, M. 1971. *Internal Labour Markets and Manpower Analysis*. Lexington: Health and Co.

Dossani, R. and Kenney, M. 2007. The next wave of globalization: relocating service provision to India. *World Development* 35(5), 772–91.

Droogleever Fortuijn, E. 2003. *Onderwijsbeleid: Maatschappelijke Functies en Strategische Keuzen.* Amsterdam: Aksant.

Felker, G. 2003. Southeast Asian industrialisation and the changing global production system. *Third World Quarterly* 24(2), 255–82.

Fields, G. 2007. *Dual Economy.* Available at: http://digitalcommons.ilr.cornell.edu/workingpapers/17/ [Accessed: 15.04.07].

FNV 2005. *Het verdwenen werk.* Amsterdam: Stichting FNV Pers, 36.

Freeman, R. 2005. Does globalisation of the scientific/engineering workforce threaten US economic leadership? *NBER Working Paper* 11457, June.

Friedman, T.L. 2006. *The World is Flat. A Brief History of the Twenty-first Century.* NewYork: Farrar, Strauss & Giroux.

Fröbel, F., Heinrichs, J. and Kreye, O. 1980. *The New International Division of Labour.* Cambridge: Cambridge University Press.

Glucksmann, M. 2004. Call configurations: varieties of call centre and division of labour. *Work, Employment and Society* 18, 795–811.

Gray, J. and Chapman, R. 2004. The significance of segmentation for institutionalist theory and public policy, in *The Institutionalist Tradition in Labor Economics*, edited by D. Chapman and J. Knoedler. New York: Armonk, 117–30.

Hudson, R. 2000. *Production, Places and Environment: Changing Perspectives in Economic Geography.* London: Prentice Hall.

Hudson, R. 2005. Rethinking change in old industrial regions: reflecting on the experiences of North East England. *Environment and Planning A* 37, 581–96.

ICFTU 2006. Surviving without quotas: a Philippine case-study. Available at: http://www.icftu.org/www/PDF/RP_Garment_Report_170605edited_260605EN.pdf [Accessed: 17.10.07].

Jetro 2008. *Philippine IT Industry Update*. Available at: http://www.jetro.go.jp/philippines/tp/itupdate_2008.html.

Krishna, A. and Nederveen Pieterse, J. 2008. The dollar economy and the rupee economy. *Development and Change* 39(2), 219–37.

Lall, S. 2000. *Export Performance and Competitiveness in the Philippines*. Oxford: QEH Working Paper, no. 49.

Lier, D. 2007. Places of work, scales of organising: a review essay of Labour Geography. *Geography Compass* 2, 814–32.

Magtibay-Ramos, N., Estrada, G. and Felipe, J. 2008. An input-output analysis of the Philippine BPO industry. *Asian-Pacific Economic Literature* 22(1), 41–56.

Mani, S. 2002. *Moving Up or Going Down the Value Chain: An Examination of the Role of Government with Respect to Promoting Technological Development in the Philippines*. Discussion paper series, 2002–10. Maastricht: UNU-INTECH.

McKay, S. 2006. *Satanic Mills or Silicon Islands? The Politics of High-tech Production in the Philippines.* Ithaca: Cornell University Press.

Peck, J. 1996. *Work-Place: The Social Regulation of Labor Markets.* New York: Guilford Press.

Peck, J. 2003. Fuzzy old world: a response to Markusen. *Regional Studies* 37(6–7), 729–40.

Perrons, D. 2007. The new economy and earnings inequalities: explaining social, spatial and gender divisions in the UK and London, in *Geographies of the New Economy*, edited by P. Daniels, A. Leyshon, M. Bradshaw, and J. Beaverstock. London: Routledge, 111–31.

Ris, L. 2007. *The Impact of No Work: Displaced Workers Livelihoods in the Philippines*. Unpublished Master's thesis, University van Amsterdam.

Rodolfo, C. 2006. *Expanding RP-US Linkages in Business Process Outsourcing,* Manila: Philippine Institute for Development Studies, Discussion paper 2006–10.

Scott, A. 2005. The shoe industry of Marikina city, Philippines: a developing country cluster in crisis. *Kasarinlan: Philippine Journal of Third World Studies* 20(2), 76–99.

Scott, A. 2006. The changing global geography of low-technology, labor-intensive industry: clothing, footwear, and furniture. *World Development* 34(9), 1517–36.

Shameen, A. 2006. The Philippines' awesome outsourcing opportunity. *BusinessWeek*, 19 September 2006.

The Economist 2007. Globalisation and the rise of inequality, *The Economist* 20 January.

Wetenschappelijke Raad voor het Regeringsbeleid 2007. *Investeren in Werkzekerheid.* Amsterdam: Amsterdam University Press.

Chapter 16
'We Order 20 Bodies'. Labour Hire and Alienation

Sylvi B. Endresen

Introduction

The title of this chapter is provided by a Namibian businessman, using the services of a labour hire agency to get a job done:

> We order 20 bodies for a month and the broker must make sure that 20 bodies come to work. It does not matter if it is the same people or not.[1]

It matters whether people are permanent or temporary, full-time or part-time workers. It matters to the capitalist, it matters to the worker. But *how* does it matter? What is the difference between permanent workers and temporary workers? And, within the latter category, temporary workers can be engaged by the capitalist directly, or made available to the capitalist through labour hire agencies. Labour hire workers share a characteristic with temporary and permanent workers; they sell their labour power to capitalists. But they do so by means of an intermediate in the labour market, the labour hire agency. Does it matter? Explanations of contemporary processes of flexibilization of work educate us on the reasons why capitalists increasingly prefer agencies, and provide explanations of the growth of contingent work and give empirical evidence of the effects of these developments upon work regimes (see GOTSU, The Geographies of Temporary Staffing Unit at the University of Manchester[2]). However, to fathom the effects upon the agents involved in the buying and selling of labour power, we need to dig more deeply into other bodies of theory that can explain effects upon human dignity when ties between employer and employee weaken. The dignity of sellers as well as buyers of labour power is at stake: people losing their humanity and turned into cynics, people reduced to mere physical labour power. I take the above citation as the

1 Labour Resource and Research Institute (LaRRI) in Windhoek has tirelessly focused on economic, social and legal aspects of labour hire. I contributed to Labour Resource and Research Institute (2002); the quote is from Labour Resource and Research Institute (2006). I am greatly indebted to Herbert Jauch and other researchers of LaRRI for my understanding of labour hire and for empirical foundations of this chapter.

2 www.sed.manchester.ac.uk/geography/research/gotsu.

point of departure, seeking answers to two questions: why did he say this, and how is it possible to say a thing like this? The quote may represent an extreme, rarely to be heard elsewhere. But the statement of this particular employer reflecting his relationship to his temporary workers, or rather, reflecting the way he chose to present it to a researcher, provides me with what I need to advocate a re-reading of Marx's theory of *alienation*. However, we need more than explanations of the dehumanizing aspects of capitalism to understand how workers can be seen as 'no more than a cipher reduced to an abstract quantity, a mechanised and rationalised tool' (Lukács 1923: 166). The (cognitive) conversion of human beings into 'things' that appear as commodities on the market is the *reification* of human relations. It follows from the extension of universal saleability, the turning of everything into commodities. Fragmentation of society into isolated individuals that pursue limited, particularistic aims is the depressing result (Mészáros 1979).

The main objective of this chapter is to draw attention to a process of purification of the commodity form of labour power involved in casualization of workers. In transactions between client companies demanding labour power and labour market intermediates supplying workers, labour power may be seen by the parties just like any other commodity. Consequently,

> It is not for the commodity to decide where it should be offered for sale, to what purpose it should be used, at what price it should be allowed to change hands, and in what manner it should be consumed or destroyed. (Polanyi 1944: 185)

Polanyi (1944: 76) considered the commodity description of labour 'entirely fictitious', a proposition that entered Labour Geography recently: Being embedded in conscious, sentient human beings, labour power should be considered a pseudo-commodity (Castree et al. 2004: 29). This chapter is a contribution to this debate.

Furthermore, another objective is to identify the sub-processes involved in labour hire; exchange of labour power, externalization and mediation; to understand how class processes are affected. The re-entry of strong commercial capital, 'merchants of labour' (Kuptsch 2006) is thus seen to characterize contemporary capitalism.

The Quote and its Context

The capitalist who referred to workers as 'bodies' undertakes business in post-Apartheid Namibia. His attitude may be dismissed as an ideological remnant: has the progress regarding workers' position in society, evident in the country's strict Labour Act, passed his house unnoticed? Labour hire work resembles contract labour arrangements that kept black labour flowing from impoverished bantustans to more prosperous white areas for half a century (Bauer 1997, Kaakunga 1990, Mbuende 1986, Moorsom 1977). It can be reflected upon in historical terms but should be understood as labour market effects of contemporary competitive pressures. Labour hire, capitalists' use of employment agencies to man the jobs is on the increase world-wide (Kuptsch 2006). Namibia is no exception: In a few years, ten agencies were established (Labour Resource and Research Institute 2006). The competitive situation of our businessman makes him go for a particular kind of work contract: Workers employed on the basis of job insecurity (no work – no pay) with low wages and few benefits (Labour Resource and Research Institute 2002, 2006). The Namibian Labour Act advocates permanent, secure jobs, decent working conditions, living wages, a union-based labour regime and state protection of workers' rights and state survival safety nets. This is still the ideal of organized labour, and indeed what all Namibian workers strive to obtain. The embryo of such a welfare regime could be observed during the first years after Independence in 1990, before globalization and the neoliberal development strategy led to a brutalization of the labour regime (Jauch and Bergene, Chapter 10, Magnusson et al., Chapter 13). The spread of the flexible labour regime is thus closely linked to the country's EPZ policy; the first agencies popped up close to the first zone (Labour Resource and Research Institute 2002). When labour hire workers are introduced, they either replace permanent or temporary workers, or result from a capitalist's choice of employees when new production is established or when his business expands.

> The presence of labour hire workers puts a downward pressure on conditions of employment for permanent workers because some companies retrench permanent workers and re-hire them through labour hire at lower rates. Salaries are cut in half and benefits are reduced. (Union organizer, Labour Resource and Research Institute 2006)

In addition to flexibility advantages, labour hire can be an effective means of controlling labour and breaking unions:

> The contract that I have signed with the client is very clear. They [the clients] don't need people who are troublemakers. They will call me to replace him with someone else. (Labour hire agency manager, Labour Resource and Research Institute 2006)

In the Namibian case, the growth of the pool of workers in the service of labour hire agencies should be understood in terms of high rates of unemployment. Workers who fail to enter the privileged segments of the labour market in desperation turn to labour hire agencies that retain between 15 and 55 per cent of their hourly rates (Labour Resource and Research Institute 2006).

Fusion of Horizons

While doing labour research in post-Apartheid Namibia I have come across a few informants who expressed racist attitudes during interviews, and if I interpret body language correctly, they *wanted* me to turn away in disgust. By increasing the distance to me, the white liberal who does not understand African reality, their standpoint is reinforced. The researcher's reaction of contempt also reveals itself in body language, reinforcing differences and making information exchange and understanding difficult. The businessman who referred to workers as bodies should be taken seriously: '[W]ords and the meaning of words are not something we construct, they are mediators of a living tradition' (Alvesson and Sköldberg 2000: 85). He revealed attitudes that put him in a bad light, and my hypothesis is that this is what he intended. To exploit other people a quantum of cynicism is needed and this quantum must be 'produced'. Others must become 'things' in our eyes, less valuable and thus separate from us. Is our businessman trying to convince himself that the labour hire workers are not *sentient beings*? Intentions are however inscrutable.

I tried not to distance myself from the informant who tried to dehumanize workers. But why should I care about a man whose views I resent? The circumstances of this businessman's life are not entirely of his own making: 'There, but for the grace of God, go I.'[3] Our values are shaped in an interaction between ourselves and social structures. Deliberate attempts at understanding the meaning-fields of others by clarifying the contrast to one's own may be termed *fusion of horizons* (Gadamer in Alvesson and Sköldberg 2000: 84), and is a fundament for empathy. Anyone who loses their humanness in any system due to properties of that particular system needs our understanding, our pity. People are seldom born with a hardened heart but may have to work hard to achieve it. In Hochschild's (2003) terms, the informant undertook 'emotional labour'; adjusting his emotions, adapting them to commercial purposes:

> Men are estranged from one another as each secretly tries to make an instrument of the other, and in time a full circle is made; one makes an instrument of himself, and is estranged from It also. (C. Wright Mills in Hochschild 2003: 24)

Not to lose our humanness, we need to recognize that persons undertaking evil acts also suffer. Although they may lack consciousness about their deprivation, they suffer a loss of dignity. This does NOT add up to excuses for heartless capitalists that mistreat workers; quite the opposite: They must be considered *human* to make them responsible for their actions; monsters are by definition irresponsible. Other Namibian employers express totally different attitudes towards labour hire workers. The director of Namport in Walvis Bay and the director of ABB during the construction of the power lines between South Africa and Namibia were against the use of casuals, especially, labour hire workers, claiming that some labour hire

3 Paraphrase of 'There but for the grace of God, goes John Bradford.' Imprisoned in the Tower of London, Bradford (1510–1555) witnessed a man going to his execution (Hendrickson 1997).

agencies treated workers 'like cattle'[4]. The analogy is not very accurate. Cattle are a valuable commodity that is not so easily replaced as workers are, and must therefore be treated well. Says Willie Eaton, a character in *The Grapes of Wrath*:

> If a fella owns a team of horses, he don't raise no hell if he got to feed'em when they ain't workin'. But if a fella got men workin' for him, he jus' don't give a damn. Horses is a hell of a lot more worth than men. (Steinbeck 1939: 374)

Should labour hire in the Namibian worst-case be considered a return to *slavery*? No. A slave that you own, you need to keep in good shape, provide food and shelter to him and his family since his well-being is important to the bottom line. We place labour hire workers firmly in the capitalist mode of production, where labour power is a commodity to be bought and sold at a marketplace where the shape, form or health of the body that contains the labour power is not the responsibility of the capitalist, but that of the individual worker. Responsibility for reproduction is, in the best case, shared with society at large.

To be Rationally Fragmented

Reification, turning persons into 'things', is an old philosophical problem. To Aristotle, the slave is not a man, but a mere thing, a 'talking tool' (Mészáros 1979: 39). Later, condensing a few centuries, in feudal society, some things were inalienable, such as human beings. For capitalism to take root such taboos had to vanish and *universal saleability* accepted:

> The *living* person, however, first had to be *reified* – converted into a thing, into a mere piece of property for the duration of the contract – before it could be mastered by its new owner. (Mészáros 1979: 34–5)

Capital needed labour free of bonds to the land, and labour was given an offer that could not be refused; to survive by selling labour power for money:

> The worker sells his labor-power in order to acquire the means of subsistence to live. To keep from dying the worker sells his life. While the capitalist buys the worker's labor-power in order to make a profit. (Ollman 1976: 170–71)

Free contractual relations were glorified as 'liberty', safeguarding individual freedom (Mészáros 1979: 255). Kant discussed such contractual arrangements in terms of *Verdingung* (conversion into a thing), but found that it was 'not a mere reification but the transference – by means of hiring out – of one's person' (Kant in Mészáros

4 Interviews undertaken during fieldwork for a study of bush encroachment in Namibia (Endresen and Wessel Pettersen 2003).

1979: 34). To me, 'contractually safeguarded freedom' sounds like an ad for Africa Labour Hire or Adecco. Liberty, however, implies the option of *not* signing; what we observe in the case of Namibian labour hire is rather 'the *contractual abdication* of human freedom' (Mészáros 1979: 34, emphasis in the original).

The cognitive construct of human things, 'bodies' treated as the instruments of capitalist production, manifests reification and is reflected in the mindset of sellers as well as buyers:

> The job title of one of the purchasing executives responsible for dealing with labour agencies is chillingly appropriate: 'Buyer, Human Resources'. (Purchell and Purchell 1998: 51–2)

Lukács develops the concept of reification of human relations; manifest when relations between people 'takes on the character of a thing, and thus acquire a "phantom objectivity"' (Lukács 1923: 83). Reification, and commodity fetishism, result when all needs are sought satisfied in terms of commodity exchange, when everything becomes universally saleable.[5]

In Marx's theory of alienation, reification, mechanization and dehumanization are linked. Starting with the capitalist labour process, *performance in time* is what matters: 'Time is everything, man is nothing; he is at the most the incarnation of time' (Marx 1847: unnumbered). Specialization and rationalization freezes time into a delimited, quantifiable continuum filled with quantifiable 'things', objects of production that are fragmented in the production process:

> This fragmentation of the object of production necessarily entails the fragmentation of its subject. In consequence of the rationalisation of the work-process the human qualities and idiosyncrasies of the worker appear increasingly as *mere sources of error* when contrasted with these abstract special laws functioning according to rational predictions. (Lukács 1923: 89)

The labour process is a mechanically objectified 'performance' whereby the worker is 'wholly separated from his total human personality' and integrated as 'a pure, naked object into the production process' (Lukács 1923: 90, 168). The objectification of the worker's labour-power into something opposed to his total personality follows from his sale of labour power as a commodity. It becomes the reality of his daily life;

> [T]he personality can do no more than look on helplessly while its own existence is reduced to an isolated particle and fed into an alien system. (Lukács 1923: 90).

5 The concept of 'commodification' is tempting here; but I suspect that this is a 'chaotic conception' (Sayer 1998: 123). I have seen it used in the sense of homogenization of the cultural world; a general acceptance of capitalist market logic; increased exchange relations; market penetration and primitive accumulation.

Capitalist production is scientific, mechanical, fragmented and specialized (Braverman 1974), the worker must likewise be 'rationally fragmented' (Lukács 1923: 90). Accordingly, the labour hire worker can by a capitalist be considered merely a body containing labour power. *Any* body.

Labour Hire and Class Processes

When labour hire increases in importance, securing flexibility is a probable hypothesis as to why: The advantage of employing only a smaller core of workers and draw upon an external large periphery when needed can be harvested by capitalists. Also, labour control by inhibiting organization may be a cause, as well advantages of externalizing the responsibility of workers in production as well as their reproduction:

> I think the main reason for labour brokering is the abdication of management responsibility. Companies ask why must we have labour hassles, why must we deal with unions, why must we deal with the Labour Act? (Labour hire agency manager, Labour Resource and Research Institute 2006)

The contemporary proliferation of labour hire agencies indicates a need for increased circulation of labour power, deepening of exploitation and increased control of labour: 'Increasingly the labour market is a site for the exchange of labour power as a purer commodity form' (Wills 2008: unnumbered).

Three related sub-process of labour hire may be identified; exchange of labour power, externalization, and mediation, all ruled by market logic. The price of labour hire power is subjected to competition between different agencies in the same market, and thus determined by supply, demand and regulation. Seen from the vantage point of the worker, he exchanges his labour power for money by means of a mediator between the capitalist at the workplace and himself. From the vantage point of the capitalist, labour hire involves the externalization of an aspect of the company's human resource management. Recruitment and part of the in-house management of some workers are no longer undertaken in-house, it is subcontracted to the labour hire agency. The labour hire agencies are the mediators that facilitate the two above processes.

Introducing the concepts of Ruccio et al. (1991) may aid the understanding of the difference between employing workers directly and employing them through a mediator. Building on Marx's value theory, they distinguish between fundamental class processes and subsumed class processes. The former takes place at the industrial site and comprises the formation of capitalist surplus-value and its appropriation by the industrial capitalist. Another name for it is exploitation. In subsumed class processes the issue is not exploitation, but 'distributions of surplus-value (to landlords, merchant capitalists, money capitalists, etc.) in order to secure some of the conditions of existence of capitalist exploitation' (Ruccio et al. 1991: 32). The industrial capitalist distributes part of the surplus-value to a commercial capitalist who markets the produce, which he buys for less than its value, and sells it for its value. The merchant is a go-between 'who compares money-prices and pockets the difference' (Marx 1894: 447). The subsumed class process secures the reproduction of the fundamental class process, facilitating capital accumulation. Where producers transfer part of the surplus-value to merchants in order to realise value faster, the relation is termed subsumed class *payment*. Whereas subsumed class payment in the theory of Ruccio et al. (1991) concerns payment for goods, in the case of labour hire it is payment for labour power. The industrial capitalist pays the commercial capitalist to supply labour power that is fed into the fundamental class process. Permanent and labour hire workers work side by side at the industrial site; they are often indistinguishable in the labour process. Whereas the remuneration of permanent workers is direct, since the industrial capitalist pays the workers' wages, that of the labour hire workers is indirect, since the industrial capitalist transfers value to the commercial capitalist who pays the wage and out of the surplus-value charges a fee for the job. The labour hire workers' wages are still part of the value they themselves create at the industrial site where they do the job. There is no fundamental class process within the labour hire agencies that involves the workers hired out. This value is transferred, in addition to some of the surplus-value appropriated by the industrial capitalist, to the labour hire agency. There is thus a difference between subsumed class payments where goods are paid for and those which pay for labour power, since the latter payment consists of surplus-value *and* wage. In the accounts, the flow may well appear as payment for *services* provided by labour hire agencies, the service being externalized management of labour. The contracts regulating these processes shift from work contracts between industrial capitalists and workers to commercial contracts between companies in which the distribution of surplus-value between industrial and commercial capitalist is negotiated. The contract between labour hire workers and the agency may have the appearance of a work contract, specifying for instance hourly rates, but can be understood as a commercial relation in which the worker surrenders part of the value he creates in return for the service of hiring out his labour power. From this fee the profit of the labour hire agency stems. Labour hire agencies are companies that profit upon handling the commodity labour power; they are 'merchants of labour' (Kuptsch 2006) with the characteristics of commercial capital:

> Commercial capital…is simply the mediating movement between extremes it does not dominate and preconditions it does not create. (Marx 1894: 447)

However, since the days of Marx the capability of commercial and financial capital to shape the conditions for their reproduction has improved. Brand name owners' control over production in factories owned by others (Wills and Hale 2005) and retailers' control over manufacturers (Hale 2005) are cases in point. Labour hire represents labour power under the immediate control of commercial capital.[6] The ability of commercial capital to shape conditions for reproduction depends on their market control: Imagine, just for a second, that *all* labour power was controlled by labour hire agencies. The agencies would still be go-betweens; they would still not undertake any production. But imagine how they would dominate.

Profit upon Alienation

To the classical economists like Steuart, the concept of alienation merely means a shift in property rights. Marx described the profit of merchants as 'profit upon alienation':

> [T]he merchant's profit is firstly made by acts simply within the process of circulation, i.e. the two acts of purchase and sale. Secondly, it is realized in the final act, the sale. It is thus 'profit upon alienation.' (Marx 1894: 447)

The concept of alienation acquired negative connotations as it came to depict negative consequences of disruptive social processes post the industrial revolution, as capitalism deepened. To the workers, this *Great Transformation* implies 'the smashing up of social structures in order to extract the element of labor' (Polanyi 1944: 172). Polanyi contrasts the condition of modern man to societies permeated by what Durkheim termed 'organic solidarity' and establishes a linkage between the conquest of the market and fundamental social change:

> To separate labor from other activities of life and to subject it to the laws of the market was to annihilate all forms of existence and to replace them by a different type of organization, an atomistic and individualistic one. (Polanyi 1944: 171)

Initially communally owned means of production were privatized during various stages of primitive accumulation during the transition to capitalism; alienation was by Marx considered one outcome. He recognized several aspects of alienation; the estrangement of man from himself, from other humans and from mankind. He saw the estrangement of man from mankind in general as 'the alienation of "humanness" in the course of its debasement through capitalistic processes' (Mészáros 1979: 15):

6 Transnational corporations that own only brand names but still control production by influencing the supply chain are 'hidden employers', and *approaches* merchant capital according to Wills and Hale (2005). Conceptual confusion may result if these externalization processes are conflated in the concept of 'outsourcing'. The brand name form should be termed 'outsourcing of jobs', labour hire is 'outsourcing of workers'. The two forms represent different strategic options for companies.

> What applies to man's relation to his work, to the product of his labour and to himself, also holds of man's relation to the other man, and to the other man's labour and object of labour. (Marx in Mészáros 1979: 15)

Rooted in the initial separation of wage workers from the means of production, 'alienation of labour' was considered by Marx the most fundamental form; turning labour power into a commodity controlled by capital thus causes all forms of alienation. Alienation under capitalism is therefore the manifestation of the antagonism between private appropriation of surplus and the social character of work (Mészáros 1979, Ollman 1976). Yet, being ingrained in human nature, the social character of work persists. In production, due to their structural positioning, capitalists and workers are strongly interdependent (Mann 1973); 'captives of each other' (Storper and Walker 1989: 168). To fully appreciate the significance of the phenomenon of labour hire, we must analyse what happens when one of the captives, the capitalist, tries to shake loose from the other while *keeping* his structural position. The explanation should be searched for in the characteristics of labour power as a commodity. Unlike commodities proper, labour power is embodied in workers who sell it for a limited period of time, they enter into social relationships and they have agency (Storper and Walker 1989); labour power is thus a pseudo-commodity (Castree et al. 2004: 29). Labour hire is a way for capital to escape two of these characteristics, social relations such as closeness, and agency such as resistance, while keeping the least problematic and most essential, the embodied labour power. Since the 'messy' human body is susceptible to emotional and physical breakdowns (Wills 2008), shifting social security nets to the state and human resource management to other companies are tempting:

> A combination of subcontracted employment relations and the exploitation of alternative supplies of workers, make it easier to extract labour power without engaging in the complications of life. (Wills 2008: unnumbered)

The capitalist can claim that the main responsibility of labour hire workers rests with the labour hire agency which is their formal employer: These workers are not on *his* hands anymore. And as we learned during research on labour hire in Namibia, workers sometimes do not know who their employer really is, who is responsible for them, the boss at the workplace, or the labour hire boss (Labour Resource and Research Institute 2002).

Let us return to the businessman who claimed that 'it does not matter if it is the same people or not' that do the job. I believe it does matter to workers making a living and searching for a meaningful life through work. Workers depend not only on the capitalists, but also on each other for their jobs as (collaborating) producers and as consumers, by Castree et al. (2004: 35) termed 'inter-worker dependency'. Employees often feel attached to their workplace; identifying with ones' work should not be alien to academics reading these lines. But when your workplace continuously shifts from one site to the other, within months, weeks, or even within a day, identification becomes difficult. For one, what would be the point of bonding with your workmates? Human dignity is at stake:

> For the alleged commodity 'labor power' cannot be shoved about, used indiscriminately, or even left unused, without affecting also the human individual who happens to be the bearer of this peculiar commodity. In disposing of a man's labor power the system would, incidentally, dispose of the physical, psychological, and moral entity 'man' attached to that tag. (Polanyi 1944: 76)

Attachment matters to workers. But does it matter to the capitalist? Employers as well as workers suffer, but capitalists may live well under alienation:

> The property-owning class and the class of the proletariat represent the same human self-alienation. But the former feels at home in this self-alienation and feels itself confirmed by it; it recognises alienation as its own instrument and in it possesses the semblance of a human existence. The latter feels itself destroyed by this alienation and sees in it its own impotence and the reality of an inhuman existence. (Marx paraphrased by Lukács 1923: 149)

When the worker 'is turned into a commodity and reduced to a mere quantity' (Lukács 1923: 166) the very same process dehumanizes the capitalist. But our ability to live in denial of deprivation varies; it is easier when the effects can be (literally) cushioned.

Conclusion

By making use of labour hire agencies, the industrial capitalist can disentangle himself from the responsibilities involved in employing workers directly. In the process he secures flexible production and supply of workers that are less likely to form strong labour unions. The labour hire worker renounces his right to negotiate contract terms with the capitalist directly. Differences in profitability between branches that might have been part of negotiating terms reduce in importance. Industrial and commercial capitalists sort out the problems of labour supply and control through commercial contracts.

I have sought inspiration in the classics; but newer theories and perspectives are needed to fully comprehend these contemporary phenomena and processes. Increased importance of labour hire can be analysed in terms of a post-Fordist rescaling process (Swyngedouw 1997), an attempt at returning to decentralized forms of capital-labour regulation (Bergene 2010). Furthermore, a characteristic of post-Fordism is accelerated labour exploitation (Peck and Tickell 1994). The growth of contingent work is a result of shifting accumulation strategies of capital, and can be considered an attack on the benefits inherent in permanent, more secure jobs. In other words, some workers are dispossessed of the potential advantages of permanent jobs. Harvey's (2003) concept of accumulation by dispossession can thus be applied to link labour hire to related phenomena of contemporary capitalism. Capital now extracts labour power in a purer commodity form (Wills 2008), creating a playground for commercial capital. Labour hire increases the magnitude of subsumed class payments, resulting in a shift of value from the sphere of production to the sphere of exchange. The increased market power of commercial capital that results should be further theorized; including the theme of this chapter, the increased control of labour by commercial capital that loosens bonds between industrial capitalists and workers. However, it is no matter of course that inhuman labour regimes should result from this process. But I do believe that labour hire makes the workers even more dependent upon a protective regulatory environment that can remove them from 'the orbit of the market' (Polanyi 1944: 186). The Namibian High Court decision is a case in point; all Labour Hire agencies should close shop by March 2009. This is a great victory of Namibian workers and Labour Resource and Research Institute in Windhoek. Scientific work matters.

It has been my contention in this chapter that labour hire represents a purification of the commodity form of labour power, underlined in Bergene's comment to this manuscript: 'Labour hire is perfecting primitive accumulation as a process; it is the fulfilment of the introduction of capitalist logic'[7]. The most precise understanding of the Namibian businessman's statement that started my journey into the theories of alienation is found in Heidegger's (quoted in Harvey 1996: 300) analysis of modern capitalism:

> In self-assertive production, the humanness of man and the thingness of things dissolve into the calculated market value.

Labour hire turns the wheels of alienation yet another turn. Will there be, as stated by Marx and Engels (1848: 222), 'left no other nexus between man and man than naked selfinterest, than callous "cash payment"'?

References

Alvesson, M. and Sköldberg, K. 2000. *Reflexive Methodology. New Vistas for Qualitative Research*. London: SAGE Publications.

Bauer, G. 1997. Labour relations in occupied Namibia, in *Continuity and Change. Labour Relations in Independent Namibia*, edited by G. Klerck et al. Gamsberg Macmillan.

Bergene, A.C. 2010. *Preaching in the Desert or Looking at the Stars? A Comparative Study of Trade Union Strategies in the Auto, Garment and Transport Industries*. Doctoral dissertation. Department of Sociology and Human Geography, University of Oslo.

7 Ann Cecilie Bergene has written Chapter 6 of this volume.

Braverman, H. 1974. *Labor and Monopoly Capital. The Degradation of Work in the Twentieth Century*. New York: Monthly Review Press, 1998.

Castree, N., Coe, N., Ward, K. and Samers, M. 2004. *Spaces of Work: Global Capitalism and the Geographies of Labour*. London: SAGE.

Endresen, S.B. and Wessel Pettersen, S. 2003. *Bush Encroachment in Namibia. A Study of Commercial Farms*. Occasional Paper 38, Samfunnsgeografi. Department of Sociology and Human Geography, University of Oslo.

Hale, A. 2005. Organising and networking in support of garment workers: why we researched subcontracting chains, in *Threads of Labour: Garment Industry Supply Chains from the Workers' Perspective*, edited by A. Hale and J. Wills. Oxford: Blackwell.

Harvey, D. 1996. *Justice, Nature and the Geography of Difference*. Cambridge, Mass.: Blackwell.

Harvey, D. 2003. *The New Imperialism*. Oxford: Oxford University Press.

Hendrickson, R. 1997. *Encyclopedia of Word and Phrase Origins*. New York: Facts on File.

Hochschild, A.R. 2003. *The Managed Heart. Commercialization of Human Feeling*. Berkeley: University of California Press.

Kaakunga, E. 1990. *Problems of Capitalist Development in Namibia. The Dialectics of Progress and Destruction*. Åbo: Åbo Academy Press.

Kuptsch, C. (ed.), 2006. *Merchants of Labour.* Geneva: International Institute for Labour Studies, ILO.

Labour Resource and Research Institute 2002. *Labour Hire in Namibia: New Flexibility or a New Form of Slavery?* Windhoek, Namibia: LaRRI.

Labour Resource and Research Institute 2006. *Labour Hire in Namibia: Current Practices and Effects*. Windhoek, Namibia: LaRRI.

Lukács, G. 1923. *History and Class Consciousness. Studies in Marxist Dialectics*. London: Merlin Press, 1990.

Mann, M. 1973. *Consciousness and Action among the Western Working Class*. London: Macmillan.

Marx, K. 1847. *The Poverty of Philosophy.* Available at: http://www.marxists.org.

Marx, K. 1894. *Capital. A Critique of Political Economy. Volume Three.* London: Penguin Books, 1991.

Marx, K. and Engels, F. 1848. *The Communist Manifesto*. London: Penguin, 2002.

Mbuende, K. 1986. *Namibia, the Broken Shield: Anatomy of Imperialism and Revolution*. Lund: Liber.

Mészáros, I. 1979. *Marx's Theory of Alienation*. London: Merlin Press.

Moorsom, R. 1977. Underdevelopment, contract labour and worker consciousness in Namibia, 1915–72. *Journal of Southern African Studies* 4(1, October), 52–87.

Ollman, B. 1976. *Alienation. Marx's Conception of Man in Capitalist Society*. Second Edition. Cambridge: Cambridge University Press.

Peck, J. and Tickell, A. 1994. Searching for a new institutional fix: the after-fordist crisis and the global-local disorder, in *Post-Fordism: A Reader*, edited by A. Amin. Oxford: Blackwell.

Polanyi, K. 1944. *The Great Transformation. The Political and Economic Origins of Our Time*. Boston: Beacon Press, 2001.

Purcell, K. and Purcell, J. 1998. In-sourcing, outsourcing and the growth of contingent labour as evidence of flexible employment strategies. *European Journal of Work and Organizational Psychology* 7(1), 39–59.

Ruccio, D., Resnick, S., and Wolff, R. 1991. Class beyond the nation-state. *Capital & Class,* 43(Spring).

Sayer, A. 1998. Abstraction. A realist interpretation, in *Critical Realism. Essential Readings*, edited by M. Archer, R. Bhaskar, A. Collier and T. Lawson. London: Routledge.

Steinbeck, J. 1939. *The Grapes of Wrath.* Penguin Classics, 2000.

Storper, M. and Walker, R. 1989. *The Capitalist Imperative.* Oxford: Blackwell.

Swyngedouw, E. 1997. Neither global nor local. "Glocalization" and the politics of scale, in *Spaces of Globalization. Reasserting the Power of the Local*, edited by K.R. Cox. London: Guilford Press.

Wills, J. 2008. *Getting over the Politics of Humpty Dumpty: Ten Propositions on Contemporary Class Formation, its Geography and its Implications for Politics*. Available at: http://www.generation-online.org/c/fc_rent12.htm.

Wills, J. and Hale, A. 2005. Threads of labour in the global garment industry, in *Threads of Labour: Garment Industry Supply Chains from the Workers' Perspectives*, edited by A. Hale and J. Wills. Oxford: Blackwell.

PART V
Conclusion

Chapter 17
Approaches to the Social and Spatial Agency of Labour

Ann Cecilie Bergene, Sylvi B. Endresen and Hege Merete Knutsen

Labour Geography is, as observed by Castree (2007), 'a work in progress', and this book is meant to be a contribution to the formidable task of systematizing and analysing labour. It is also about the *consolidation* of approaches in Labour Geography. Consolidation involves not disregarding the insights from already existing literature but seeking new and innovative ways of combining insights gained in a variety of fields and across time. Hence, we hope to inspire Labour Geographers to explore grand theory classics when shedding light on contemporary challenges as well as to venture beyond established disciplinary boundaries.

The aim of this particular chapter is to insert the insights developed by the contributors to this volume in the on-going discussion of what issues Labour Geographers need to deal with. The book features theoretical discussions on agency, union renewal and the analytical merit of the concept of labour regimes. Furthermore, it raises some important questions related to space and spatial fixes, the challenges for unions in going global and for us in going north-south in our case studies. We will demonstrate how the different chapters build on the theoretical foundations of Labour Geography, formulate new theoretical positions and point to the need for new empirical and theoretical explorations. In the following discussion we will actively relate to the suggestions made by Castree (2007), Lier (2007), Coe and Jordhus-Lier (Chapter 3), and to the lacunas in Labour Geography identified by Herod (Chapter 2).

Agency of Labour

A major contribution of Labour Geography to the wider field of Human Geography is its emphasis on the agency of workers. Despite this signature characteristic, agency is still under-theorized and under-specified (Castree 2007, Herod, Chapter 2). In Castree's view, the term agency has become a word merely classifying instances of worker actions described in empirical studies rather than a theoretical concept helping us understand the dynamics of disputes or actions. In most of the chapters in this volume, agency has *not* been dealt with as sudden and spectacular outbursts of militancy, but rather as part and parcel of the everyday struggles of workers and unions. Meyer and Fuchs (Chapter 8) analyse how German unions develop dynamic capabilities in their effort to organize temporary workers, while Magnusson, Knutsen and Endresen (Chapter 13) consider how labour regimes are formed not through disputes at fever pitch but rather day-to-day mechanisms of control and resistance. Furthermore, Tufts (Chapter 7) explains union agency in Schumpeterian terms, while Cumbers and Routledge (Chapter 4) and Ryland (Chapter 5) address every-day strategies of labour internationalism.

Coe and Jordhus-Lier (Chapter 3) call for a relational understanding of agency. By extension, we would argue that a distinction should be made between agent and actor. The term actor implies playing a pre-defined role in line with Archer's (1995: 277, emphasis in original) reasoning that 'we become Agents *before* we become Actors'. By employing the term agent the focus is shifted to 'the location of individuals in hierarchical social relations which furnishes them with social interests' (Creaven 2000: 177). However, the distinction is analytical, and, as Creaven recognizes, individuals are also role-incumbents, although their social agency and person-hood cannot be reduced to it. Sharing common interests with other agents by virtue of structural positioning, i.e. being part of what Creaven (2000) terms agential collectivities, individuals are either endowed with certain powers or subjected to oppression and exploitation, which they do not acquire voluntarily or as part of a role they occupy of their own choice. The point of labouring this distinction becomes clear when individuals occupying different roles still share agential interests. For example, few would deny that unemployed share certain concerns with the employed, both sharing the location of 'workers' in the capitalist social structure. Both have real interests in, at least, the availability of decent jobs paying high wages.

In a critical realist perspective, structures are regarded as 'emergent' from the activities and attitudes of previous generations, and they do not specify how individuals must think and act, but rather define 'situational logics' (Creaven 2000). Hence, agents and agency are not structurally determined, even though they are structurally conditioned. Additionally, agents interact with existing structures either through reproducing or changing them. The interplay between agents and structures can be understood as a never-ending series of cycles consisting of structural conditions, social interaction and structural development. Creaven (2000) maintains that agents reflexively monitor and appraise their situation, and against this backdrop, base their choice on normative considerations, decide what actions are appropriate to better their plight, and then act purposively in light of this analysis. Social structures and hierarchical social relations endow agents with different structural capacities and furnish them with interests which may motivate collective struggles. In other words, the location of individuals in specific structural relations vests them with specifiable interests common to whole agential collectivities, which in turn impact upon their norms and beliefs. However, save perhaps the interest in surviving, the very process of identifying and defining interests is a normative one and depends on which agential collectivity is taken into account in any particular instance (Creaven 2000). This proposition led Bergene (Chapter 6) to argue that agents often display a chaotic consciousness.

Moreover, Creaven (2000) argues that there is a need to distinguish primary agency from corporate agency. Primary agency pertains to agents that are unorganized and often react and respond to their context without any intentional motive for generating societal effects. In contrast, corporate agency refers to agents who are organized in the pursuit of certain interests and who thus engage in the articulation of these interests and in concerted action in order to bring about structural change. Consequently, corporate agency depends on some degree of consciousness, or at least awareness, on the part of agents as to which interests are worthy of, and susceptible to, collective mobilization. This expands on the more common way of distinguishing between individual and collective agency in Labour Geography (see Kelly 2002, Cumbers, Nativel and Routledge 2008), both of which are understood as intentional. In their study of union efforts to recruit temp workers, it is the notion of corporate agency that underlies both the person-oriented and organizational perspective of Meyer and Fuchs (Chapter 8). Furthermore, drawing on Herod's (Chapter 2) transfusion of Aristotle's differentiation of four types of causality into the theorization of agency, we could argue that material, efficient, formal and final causes relate to agency pertaining to organizations, subjects, strategies and overriding motivation respectively. Returning to the contribution of Meyer and Fuchs, the motivation of struggle is to better the plight of (German) workers, while the strategy of struggle is to meet the challenge of an increasing proportion of temporary workers in the workforce in Germany by organizing them. What Meyer and Fuchs term organizational perspective can be seen as agency pertaining to the organizations of struggle, while the person-oriented perspective relates to the subjects of struggle.

Workers occupy a number of structural positions, and this makes it necessary to relate their agency to their positionality and to put analyses of power back on the table (Coe and Jordhus-Lier, Chapter 3). Coe and Jordhus-Lier thus agree with Castree (2007) that further developing our understanding of agency, especially its embeddedness, is an important task if Labour Geography is to be advanced as a theoretical, methodological and even as a political project. They argue that agency needs to be re-embedded in order for us to be able to grasp the multiple structural positions occupied by workers who 'constantly operate within complex and variable landscapes of opportunity and constraint' (Chapter 3, page 39). According to Coe and Jordhus-Lier, there are three social structures that are fundamental to labour, and in relation to which the agency of labour should be embedded; capital, the state and the community. In more abstract terms, we could say that the agency of labour needs to be embedded relationally, i.e. paying heed to how the agency of labour is shaped by its relation to capital and to the state, and spatially, i.e. workers need to be understood more holistically and not simply as members of a class, but also members of neighbourhoods, communities, cities, nations and so on, all of which feed into their consciousness and shape their response to capital and the state. We will deal with the tension that might arise between class and spatial interests below.

Agency of the State and Capital

Both Castree (2007) and Herod (Chapter 2) consider the state to be under-theorized by Labour Geographers. Rather than being neutral or the instrument of one class, the state may best be thought of as a social relation reflecting the prevailing balance of social forces (Poulantzas 1976, Jessop 2008). Since social relations know no boundaries, this has implications for whether the state can still be viewed as the focal point of social transformation (Holloway 2002). Hence, the state is limited; 'just one node in a web of social relations', the scope of which reaches far beyond that of any state (Holloway 2002: unnumbered). This is demonstrated in Knutsen and Hansson (Chapter 12), studying transition economies. Although the agency of states seems all-encompassing, labour regimes cannot be understood unless the complex relations between state, capital and labour are analysed. However, as pointed out by Holloway, the social relation between capital and labour stretches far beyond the borders of any nation, necessitating an understanding of the trans-national dynamics involved. As pointed out by Kelly (2002) in his analysis of export processing zones, we need to pay heed to the different interests and power relations between foreign capital, on the one hand, and local capital and authorities on the other. While foreign companies may try to promote better working conditions among their suppliers, this is counteracted by local capital and authorities. Similarly, Magnusson, Knutsen and Endresen (Chapter 13) point out how state regulation and analyses of labour regimes at the workplace level need to be sensitive to relations above and below the national scale. For instance, they demonstrate how the neoliberal policies of the elite hamper the implementation of labour-friendly regulations.

Holloway (2002) thus argues against an instrumentalist perspective since it involves abstracting the state from the power relations in which it is embedded. Although the state is legitimized through notions of sovereignty and of universal interests, it has a class character. In this volume, the role of the state in negotiating class interests in a post-Apartheid society, where revolutionary rhetoric masks the class character of the state, is discussed by Jauch and Bergene (Chapter 10). Additionally, Doucette (Chapter 11) argues that it is not very fruitful to seek explanations of social events with reference to the capital-labour relation alone. For similar reasons, several contributors (Chapters 10, 11, 13) take Gramsci as a point of departure. Gramsci emphasized how social relations interpenetrate with relations of production, and how capitalist societies are characterized by complex networks of relations between classes and social forces (Simon 1991). One of the institutions these networks give rise to is the state, which holds monopoly over the coercive apparatus, and has thus become a major area of struggle. Doucette (Chapter 11) demonstrates this interconnectedness in a concrete analysis of democratization in an authoritarian state. Bodies of cooperation are established by the state, creating tensions that require analysis of the agency of the state.

One dimension of state agency is, as we have seen above, the state as regulator, safeguarding workers' rights – or ignoring them. Another dimension is the state as provider of infrastructure or educational services, the quality of which has long-term effects upon capital as well as labour. This forms part of the argument of Taylor and Bryson (Chapter 14). Discussing how neoliberal policies have undermined in-house training, and thus workers' skills, and how government training programmes are poor and disconnected from the real world, they claim that competitiveness has been 'eroded from below'.

The contributors to this volume have related more actively to the state than Labour Geographers have previously done, shedding light on the state as a site, generator and product of strategies in the politics of labour. Furthermore, the state is present in the analyses both as a container of labour practices; a provider of basic services; a political apparatus and an employer. The strong control that the state exerts over labour is prominently present. On the other hand, Andrae and Beckman (Chapter 9) advocate stronger state agency if working conditions among informal workers are to be upgraded, although they place weight on the role of unions in that endeavour.

As stated in Chapter 1, attention to the agency of capital is lacking in Labour Geography. This may be explained by reference to its origin as a reaction to the over-emphasis on capital at the expense of labour in Economic Geography. There is thus a need to unpack capital in order to understand how and why it operates in different ways *vis-à-vis* labour in different contexts. Hence, we should examine the interests and competitive strategies of different types of firms in different industrial sectors, analyse how they are spatially embedded, what characterizes the set of power relations that they are part of, what their position and leverage in this is, and how they operate on the basis of this position. In this effort Labour Geography may benefit further from the debate on embeddedness (see for instance Coe et al. 2004, Hess 2004), and the work on conceptualizing the firm in Economic Geography which also addresses the collective agency of firms (see for instance Taylor and Asheim 2001, Taylor 2006).

Another way of addressing the agency of capital is to explore and further the theoretical tradition of political economy. Endresen (Chapter 16) analyses the increasing importance of labour market intermediates such as temp agencies. Drawing upon classical sources such as the labour theory of value, reification and alienation, she discusses what perspectives on social reality labour hire necessitates, and how the phenomenon of labour hire affects social existence.

Space

Doucette (Chapter 11) argues that a constructive tension should be kept up between the political economy of labour and the political geography of labour. The former concerns labour as a moment in capital accumulation, the latter labour as a social and political force. Moreover, the concept of spatial fix (Harvey 2002) should not be reserved for capital migration, but also include technological, product and political fixes since they are all deeply spatialized. He thereby touches on the longstanding debate between Andrew Sayer and David Harvey on the difference that space makes (see for instance Castree 2002). Sayer (1992) argues that lest spatial fix be rendered a chaotic conception, it must isolate a necessary relation. Although both Sayer and Harvey adhere to a relational notion of space in which space is viewed as '*both* medium and outcome, as consequence *and* cause' (Castree 2002: 191, emphasis in original), and thus agree that space is neither a container nor simply an expression of social processes, they disagree on to what extent space pervades all social phenomena and thus on the possibility of abstracting from space. While Harvey has made significant attempts at developing historical-*geographical* materialism, and thus adding space to Marxist theories in a way that makes it internal and necessary to the argument, Sayer is of the opinion that it may both be necessary and permissible to abstract from space, i.e. 'theorise relations, processes and structures *as if* they had little or no spatial integument' (Castree 2002: 191, emphasis in original).

Heeding the importance of the concept of spatial fix as theorized by Harvey, Herod (1995, 1997) claims that workers are active geographical agents striving to create their own spatial fixes in order to secure their daily and generational self-reproduction. Paraphrasing Marx, Herod argues that 'whilst workers make their own geographies they do not do so under conditions of their own choosing' (Chapter 2, page 19). Furthermore, the ideal fixes for workers might conflict with those of the state, capitalists and even workers elsewhere. Hence, in Herod's view geography still matters, and labour matters to geography. Contrary to the view among hyper-globalists, Herod emphasizes that geography has not been annihilated since the reduction in spatial barriers carries with it an increasing importance of geography; the ability to be spatially intelligent and flexible becomes more and more important. This recognition carries with it the consequence that both capital and labour have different degrees of place fixedness as Herod terms it (Chapter 2) or, in Bergene's (2010) words, different degrees of dependence on particular spaces for the realization of essential interests. Furthermore, neither the state, capital nor labour are monolithic socio-spatial agents, since they are internally differentiated when it comes to the spatial fixes preferred (Herod, Chapter 2). This brings us back to the tension between class and spatial interests mentioned above.

Important for the theorization of the negotiation of class versus spatial interest is the work of Johns (1998). In her article on solidarity between US and Guatemalan workers, she distinguishes between accommodationist and transformative solidarity. While the former denotes solidarity deriving from spatial interests, being first and foremost directed at defending what Cox (1998) terms spaces of dependence, the latter denotes solidarity deriving from class interests in overthrowing capitalism regardless of short-term spatial interests. Hence, while accommodationist solidarity actions could involve campaigns directed at increasing the wages of workers elsewhere in order to reduce their competitive advantage in the eyes of capital (although under the pretext of justice), transformative solidarity actions would, in the strictest sense, involve ideologies and actions directed towards abolishing wage labour, or at least entail anti-capitalist critique and class consciousness, what Bergene (2010) has termed reformist solidarity.

The challenges of building international solidarity are treated in Cumbers and Routledge (Chapter 4), who underline the social and spatial embeddedness of agents, and point out the weaknesses of top-down models of solidarity in the union movement. They introduce the concept of 'entagled geographies' to denote the contradictions and tensions which surface when the interests of labour at different levels meet visions of international solidarity. Furthermore, workers have a life beyond being a worker, and Ryland (Chapter 5) argues that building an internationalist class consciousness is upset by unionists' and workers' understanding of the rationale of solidarity, whether affective (transfomative) or instrumental (accommodationist), but also by the multiple and fragmented identities among workers. She thus raises the question of how labour internationalism is developed through a continuous negotiation between the union leadership and union members. Similarly, Bergene (Chapter 6) argues that people's consciousness is informed by a plethora of agential interests and ideologies, giving rise to what she terms chaotic consciousness in the union movement, and suggests an engagement with Freire's Dialogical Method as a way of negotiating it.

Union Renewal

The focus in Labour Geography so far has been on scales of union organization and on new models of organizing, such as social movement unionism and community unionism (see for instance Herod 1995, 2001, 2002, 2003, Wills 1998, 2001, 2002, Castree 2000, Waterman 2001, Waterman and Wills 2001, Bergene 2007, Lier 2007, McBride and Geenwood 2009). Instead of consigning traditional unions to the dustbin, several of the contributions to this volume point to how they are trying to grapple with new challenges in innovative ways; some chapters even venture to point forward to potentials for improvement. For instance, Meyer and Fuchs (Chapter 8) introduce research on innovation in Economic Geography in their analyses of how unions organize temporary workers in Germany. Another example is Ryland's (Chapter 5) and Bergene's (Chapter 6) attempt at understanding the challenges that unions face due to the distance between the union leadership and the rank and file, both of whom point at out how the obstacles may be surmounted. By introducing the notion of Schumpeterian unions, Tufts (Chapter 7) brings research on unions forward by transcending the binary notion of business (traditional) and social movement unions. Hence, as emphasized by Herod (Chapter 2), unions should not be treated as monoliths.

A shift from focusing solely on social movement unionism back to traditional unionism can be found in Andrae and Beckman's contribution (Chapter 9). While the issue of organizing informal workers has mainly been addressed in light of social movement unionism or as NGO-driven, Andrae and Beckman argue that traditional unions, in cooperation with associations in the informal economy, may offer better access to the social welfare services of the state. However, the applicability of their model will depend on the role and leverage of traditional labour movements and the characteristics of the states in question. It is also important to theorize the potentially contradicting interests between unions and informal economy organizations, as well as how the asymmetrical power relations between the two might impact upon the effectiveness of measures that are meant to benefit workers in the informal economy. A recent study of unionization among informal workers in the Indian construction and tobacco industries shows that it is possible to organize these workers and that this effort may result in a shift in union focus from labour rights to welfare rights (Agarwala 2006, 2008).

Labour Regimes

The concept of labour regime draws on two major theoretical sources of inspiration: Braverman's (1974) understanding of control of workers at the workplace and the regulation approach which seeks to explain why capitalism, despite inherent crisis tendencies, stabilizes over long periods of time (Lipietz 1994, Aglietta 1997). Burawoy's (1985) concept of factory regime, Jonas' (1996, 2009) concept of labour control regime, and Andrae and Beckman's (1998) labour regime all draw upon these traditions.

The chapters by Knutsen and Hansson (Chapter 12) and Magnusson, Knutsen and Endresen (Chapter 13) on labour regimes reveal how the plight of workers as producers may be understood through contextualizing the agency of workers, unions, capital and the state. Working conditions, control of labour and resistance to poor working conditions have been the main issues to be explained in studies of labour regimes. However, as such analyses require attention to conditions outside the workplace, the analytical framework opens for a more holistic analysis of the lives of workers. Knutsen and Hansson (Chapter 12) touch upon this in two ways. First, they argue that studies of labour regimes may draw on welfare regime approaches when analysing institutional changes that lead to commodification and decommodifiaction of labour, which in turn affect social differentiation and the forms that labour activism takes. Second, they argue that labour regimes and political regimes be studied as interconnected processes, since political regimes condition the way that labour may respond to working conditions and general welfare.

Hence, we would argue that analyses of labour regimes may benefit from more explicit recognition of, and active engagement with, the regulation approach, as the latter theorizes the establishment, development and transformation of regimes of accumulation and modes of regulation (Lipietz 1994, Aglietta 1997). We are not suggesting an implicit employment of the regulation approach, merely adopting their conceptual framework to point out the erosion of the Fordist model of socio-economic development, i.e. the combination of a regime of accumulation and mode of regulation, and the emergence of Post-Fordism, or more flexible ways of production, in explaining the shift from a hegemonic to a more despotic labour regime (Burawoy 1985, Nichols et al. 2004). We would argue that a more systematic analysis of what changes take place in the mode of regulation is needed, with a particular focus on how this affects the macro-politics of the state pertaining to labour and how it translates at the micro-level of the workplace. Furthermore, understanding changes in regimes of accumulation may enhance our knowledge of how and why labour regimes change. Connecting labour regimes to a particular model of socio-economic development may be fruitful since the issues of importance to labour under different models of socio-economic development are likely to differ. We could, for instance, assume that while workers at a car plant in the 1950s would be struggling against repetitive tasks, autoworkers today may be resisting multitasking. This could be because job rotation might weaken union organizing efforts since workers no longer possess unique skills and are furthermore separated from each other to the extent that they might find it harder to build bonds of solidarity.

Furthermore, modes of regulation are political both in their establishment and consolidation since they are products of socio-political struggles and institutionalized compromises (Lipietz 1994). As such, social institutions must be legitimized on the basis of collective values, and the ruling class needs, in a Gramscian sense, to ensure hegemony. If successful, social struggles might be stabilized for extended periods of time. Fordism is one such stable period of accumulation and industrial peace. However, Fordism entered a terminating crisis in the 1970s, and the last couple of decades has seen debates on whether Fordism, both as an accumulation regime and a mode of regulation, has been superseded by a new model of socio-economic development, which has by some been termed post-Fordism (see for instance Jessop 1992, Peck and Tickell 1994). The transformation from Fordism to post-Fordism had grave implications for workers since labour has been pushed on the defensive both as a political discourse and as a movement. Labour parties no longer speak the language of class and unions are struggling to keep their members, let alone increase their membership. Deregulation, a post-Fordist phenomenon often regarded as weakening workers and their unions, involves the state and other social institutions (re)regulating the accumulation of capital. As Gramsci (1932–34) pointed out, *laissez-faire* is just another form of state regulation since it is introduced and maintained by legislative and coercive means. One of the main forms of reregulation after the Fordist crisis is captured by Harvey's (2003) concept of *accumulation by dispossession.* Accumulation by dispossession involves processes of deregulation, fragmentation, casualization, flexibilization, informalization and commodification of labour, all of which are new real-world developments set in motion after the crisis of Fordism. Understanding these processes is a formidable task that raises a host of new questions and challenges for Labour Geographers. One of them is the phenomenon of labour hire, which Endresen (Chapter 16) has theorized as a purification of the commodity form of labour since labour power is treated just like any other commodity and not as a pseudo-commodity embedded in conscious, sentient beings.

Going Global and Going North-South

With analyses of labour issues in countries such as South Korea, Namibia, China, Vietnam, Nigeria and the Philippines this book contributes to a geographical expansion of our analyses, but we are not shifting focus away from previous 'heartlands' such as Canada, Germany and the UK. The main advantage of going global and going north-south in our selection of themes and case studies is that we become more open to variation in power relations and more context-sensitive in our explanations of the possibilities and constraints that labour is subjected to.

Moreover, we find that it may be fruitful to combine theoretical approaches that were initially developed with an emphasis on the global north with approaches more directly developed for explanations of phenomena in the global south. Beerepoot (Chapter 15) does this when he supplements theory on labour market segmentation, which was developed on the basis of conditions in the global north, with insights from the old debate on a new international division of labour (NIDL) and newer literature on the second global shift, which focus on the global south. Implicitly, he contributes to the wider debate on the need to rethink theories that emanate from the 'heartlands', i.e. the global north, of Labour Geography in analysing changes in other parts of the world. On the one hand, mid-range theories require some grounding in the geographical areas under study. On the other hand, attempts to cross-fertilize theoretical approaches and insights from studies in different parts of the world may inspire reflection and critical questions. However, with the long-term bias in favour of the global north in Labour Geography, we may need more knowledge on what theories grounded in these 'heartlands' are *incapable* of explaining in the global south.[1] To our knowledge there has still not been much interest in exploring how insights from the global south may inform studies in the global north. For example, analyses of social partnership in the 'heartlands' may benefit from knowledge of experiences with such agreements in countries where the opposition against it has been stronger, such as the comparisons made between Namibia, Ireland, Belgium and the US by Bergene (2010).

Links Still Missing in Labour Geography

We believe that this book constitutes steps in the right direction both by way of theorizing agency as it pertains to the state and capital as well as labour, and analyzing unions and labour regimes. The state is by no means a blind spot in the contributions to this volume, although we find a *systematic* attempt at theorizing it still somewhat missing in Labour Geography. More work also remains to be done in order to develop analytical frameworks that may unpack capital and position it in relation to labour. However, a return to some of the core issues in Economic

1 Comment by an anonymous referee.

Geography, such as the debate on embeddedness and the conceptualization of the firm, may inspire us in this endeavour.

Although studies of unionization among informal workers are not totally missing anymore, more research is required in this field, particularly on the implications for workers if unions fight for improved welfare schemes instead of working conditions. Both traditional and 'new' unions might engage in the informal economy, and we would argue that more work needs to be done on what characterizes such 'new' unions and what distinguishes them from traditional ones. The analytical framework developed by Tufts (Chapter 7) may be helpful in this regard. In light of the real-world developments of casualization and decommodification of labour, the agency of the unemployed is a missing link urgently in need of being addressed along with the question of why the union movement has always been ambivalent to the 'reserve army of labour'. Another development that needs to be better understood is the formalization of this reserve army under the wings of temp agencies or labour hire companies. A main way forward in terms of analysing labour regimes is to enhance our knowledge of how and why labour regimes change, and we have argued that careful operationalization of the regulation approach may provide a good framework to attain this. However, the challenge of connecting particular labour regimes to generalized regimes of accumulation and modes of regulation remains unresolved.

The geographical and sectoral expansion of our studies has enhanced our knowledge of contextual variation and thus taught us to become more context-sensitive in our explanations. This leads us back to the seminal work done by Peck (1996) on workplaces. By combining an emphasis on place, and thus sensitivity to people's embeddedness, for instance their reproduction and participation in civil society, with a more traditional focus on production and the more general processes unfolding in space, it forestalls myopia. What might be missing here is further cross-fertilization of theories and perspectives grounded in one part of the world with those developed in other parts. However, more important in order to improve our explanations and further development of theory is the big task of making systematic comparative analyses of a wide range of case studies in context, as suggested by Castree (2007).

References

Agarwala, R. 2006. *From Worker to Welfare. Informal Workers' Organizations and the State in India.* Dissertation presented to the Faculty of Princeton University in candidacy for the degree of doctor of philosophy.

Agarwala, R. 2008. Reshaping the social contract: emerging relations between the state and informal labour in India. *Theory and Society* 37(4), 375–408.

Aglietta, M. 1997. Postface to the new edition: Capitalism at the turn of the century: regulation theory and the challenge of social change, in *A Theory of*

Capitalist Regulation. The US Experience, edited by M. Aglietta. New Edition 2000. London: Verso.

Andrae, G. and Beckman, B. 1998. *Union Power in the Nigerian Textile Industry. Labour Regime and Adjustment*. Nordiska Afrikainstiutet, Uppsala.

Archer, M. 1995. *Realist Social Theory: The Morphogenetic Approach*. Cambridge: Cambridge University Press.

Bergene, A.C. 2007. Trade unions walking the tightrope in defending workers' interests: Wielding a weapon too strong? *Labor Studies Journal* 32(2), 142–66.

Bergene, A.C. 2010. *Preaching in the Desert or Looking at the Stars? A Comparative Study of Trade Union Strategies in the Auto, Garment and Transport Industries*. Doctoral dissertation. Department of Sociology and Human Geography, University of Oslo.

Braverman, H. 1974. *Labor and Monopoly Capital. The Degradation of Work in the Twentieth Century*. Reprinted edition 1998. New York: Monthly Review Press.

Burawoy, M. 1985. *The Politics of Production. Factory Regimes Under Capitalism and Socialism*. London: Verso.

Castree, N. 2000. Geographic scale and grass-roots internationalism: the Liverpool dock dispute, 1995–1998. *Economic Geography* 76(3), 272–92.

Castree, N. 2002. From spaces of antagonism to spaces of engagement, in *Critical Realism and Marxism*, edited by A. Brown et al. London: Routledge.

Castree, N. 2007. Labour Geography: a work in progress. *International Journal of Urban and Regional Research* 31(4), 853–62.

Coe, N.M., Hess, M., Yeung H.W., Dicken, P. and Henderson, J. 2004. 'Globalizing' regional development: a global production networks perspective. *Transactions of the Institute of British Geographers* 29, 468–84.

Cox, K.R. 1998. Spaces of dependence, spaces of engagement and the politics of scale, or: looking for local politics. *Political Geography* 17(1), 1–23.

Creaven, S. 2000. *Marxism and Realism. A Materialistic Application of Realism in the Social Sciences*. London: Routledge.

Cumbers, A., Nativel, C. and Routledge, P. 2008. Labour agency and union positionalities in global production networks. *Journal of Economic Geography* 8(2), 369–87.

Gramsci, A. 1932–34. Some theoretical and practical aspects of economism, in *A Gramsci Reader. Selected Writings 1916–1935*, edited by D. Forgacs. London: Lawrence and Wishart.

Harvey, D. 2002. *The Limits to Capital*. Oxford: Basil Blackwell.

Harvey, D. 2003. *The New Imperialism*. Oxford: Oxford University Press.

Herod, A. 1995. The practice of international labor solidarity and the geography of the global economy. *Economic Geography* 71(4), 341–63.

Herod, A. 1997. From a geography of labor to a labor geography: labor's spatial fix and the geography of capitalism. *Antipode* 29(1), 1–31.

Herod, A. 2001. Labor internationalism and the contradictions of globalization: Or, why the local is sometimes still important in a global economy, in *Place, Space and the New Labour Internationalisms*, edited by P. Waterman and J. Wills. Oxford: Blackwell.

Herod, A. 2002. Organizing globally, organizing locally: union spatial strategy in a global economy, in *Global Unions? Theory and Strategies of Organized Labour in the Global Political Economy*, edited by J. Harrod and R. O'Brien. London: Routledge.

Herod, A. 2003. Geographies of labor internationalism. *Social Science History* 27(4), 501–23.

Hess, M. 2004. Spatial relationships? Towards a reconceptualization of embeddedness. *Progress in Human Geography* 28, 165–86.

Holloway, J. 2002. *Change the World Without Taking Power*. Available at: http://libcom.org/library/change-world-without-taking-power-john-holloway. [Accessed: 21.07.08].

Jessop, B. 1992. Fordism and post-Fordism. A critical reformulation, in *Pathways to Industrial and Regional Development*, edited by M. Storper and A. Scott. London: Routledge.

Jessop, B. 2008. *State Power: A Strategic-Relational Approach*. Cambridge: Polity Press.

Johns, R.A. 1998. Bridging the gap between class and space: U.S. workers solidarity with Guatemala. *Economic Geography* 74(3), 252–71.

Jonas, A.E.G. 1996. Local labour control regimes: uneven development and the social regulation of production. *Regional Studies* 30(4), 323–38.

Jonas, A.E.G. 2009. Labor control regime, in *International Encyclopedia of Human Geography*, edited by A.E.G. Jonas, R. Kitchin and N. Thrift. Amsterdam: Elsevier, 59–65.

Kelly, P.F. 2002. Spaces of labour control: comparative perspectives from Southeast Asia. *Transactions of the Institute of British Geographers* 27(4), 395–411.

Lier, D. 2007. Places of work, scales of organising: a review of Labour Geography. *Geography Compass* 1(4), 814–33.

Lipietz, A. 1994. Post-Fordism and democracy, in *Post-Fordism: A Reader*, edited by A. Amin. Oxford: Blackwell.

McBride, J. and Greenwood, I. (eds), 2009. *Community Unionism. A Comparative Analysis of Concepts and Contexts*. Hondsmills: Palgrave MacMillan.

Nichols, T., Cam, S., Wenchi, G.C., Chun, S., Zhao, W. and Weng, T. 2004. Factory regimes and the dismantling of established labour in Asia: a review of cases from large manufacturing plants in China, South Korea and Taiwan. *Work, Employment and Society* 18(4), 663–85.

Peck, J. and Tickell, A. 1994. Searching for a new institutional fix: the after-Fordist crisis and the global-local disorder, in *Post-Fordism: A Reader*, edited by A. Amin. Oxford: Blackwell.

Peck, J. 1996. *Work-Place. The Social Regulation of Labor Markets*. New York: The Guilford Press.

Poulantzas, N. 1976. The capitalist state: a reply to Miliband and Laclau. *New Left Review* 95 (January–February), 63–83.
Sayer, A. 1992. *Method in Social Science: A Realist Approach.* Second Edition. London: Routledge.
Simon, R. 1991. *Gramsci's Political Thought: An Introduction.* Revised Edition. London: Lawrence & Wishart.
Taylor, M.J. 2006. The firm: coalitions, communities and collective agency, in *Understanding the Firm: Spatial and Organizational Dimensions*, edited by M.J. Taylor and P. Oinas. Oxford: Oxford University Press, 87–116.
Taylor, M.J. and Asheim, B.T. 2001. The concept of the firm in economic geography. *Economic Geography* 77(4), 315–28.
Waterman, P. 2001. Trade union internationalism in the age of Seattle. *Antipode* 33(3), 312–36.
Waterman, P. and Wills, J. 2001. Place, space and the new labour internationalisms: beyond the fragments?, in *Place, Space and the New Labour Internationalisms*, edited by P. Waterman and J. Wills. Oxford: Blackwell.
Wills, J. 1998. Taking on the CosmoCorps? Experiments in transnational labor organization. *Economic Geography* 74(2), 111–30.
Wills, J. 2001. Uneven geographies of capital and labour: the lessons of European Works Councils. *Antipode* 33(3), 484–509.
Wills, J. 2002. Bargaining for the space to organise in the global economy: a review of the Accor–IUF trade union rights agreement. *Review of International Political Economy* 9(4), 675–700.

Index

For Product Safety Concerns and Information please contact our EU representative GPSR@taylorandfrancis.com Taylor & Francis Verlag GmbH, Kaufingerstraße 24, 80331 München, Germany